SEPTIC–TANK SYSTEMS
A Consultant's Toolkit

Volume I
Subsurface Disposal of
Septic-tank Effluents

SEPTIC–TANK SYSTEMS
A Consultant's Toolkit

Volume I
Subsurface Disposal of
Septic-tank Effluents

by
John H. Timothy Winneberger

BUTTERWORTH PUBLISHERS
Boston • London
Sydney • Wellington • Durban • Toronto

An Ann Arbor Science Book

Ann Arbor Science is an imprint of Butterworth Publishers

Library of Congress Cataloging in Publication Data

Winneberger, John H. Timothy.
Septic tank systems.

"An Ann Arbor science book."
Includes index.
1. Septic tanks. I. Title.
TD778.W56 1983 628'.742 83-7708
ISBN 0-250-40635-7 (set)
ISBN 0-250-40586-5 (v. 1)
ISBN 0-250-40634-9 (v. 2)

10 9 8 7 6 5 4 3 2 1

Butterworth Publishers
80 Montvale Avenue
Stoneham, MA 02180

Printed in the United States of America

DEDICATION

It is very much hoped these writings find interest and beneficial applications. If so, I will feel justified in having undertaken the task of presenting them. Also, the arduous labors of Carin Winneberger will have been justified. Friends have encouraged, but Carin has been the one to successfully lead me to the task. She has devoted many years in helping me gather field data and many weeks with these writings. And, she has been responsible for translating as many of my yesteryear's expressions into modern English as my stubborness would yield. The younger readers may notice a strange word, expression, or punctuation, here and there. By the bye, it is to Carin this work is dedicated.

CONTENTS

ABOUT THE AUTHOR

John H. Timothy Winneberger, PhD, graduated from the University of Oklahoma following service in the U.S. Army Air Corps during World War II. After a brief teaching period, Dr. Winneberger completed graduate studies in plant physiology at the University of California, Berkeley, and received his PhD in 1957. Beginning with a weed control study for the Pineapple Research Institute in Honolulu, Hawaii, in 1957, Dr. Winneberger conducted a variety of research projects and published on plant physiology.

Starting in 1958 at the University of California Sanitary Engineering Research Laboratory, he directed an eight-year research program for the Federal Housing Administration on the designs, lifespans and performance characteristics of septic-tank systems. This study, the first of its kind, established Dr. Winneberger as a leader in the development of sewage disposal technology and an expert in the design of improved septic-tank systems.

Since 1966 Dr. Winneberger has been a full-time consultant. He has advised governments, corporations, health departments, real estate boards and private homeowners concerning practical solutions to sewage disposal challenges and problems. Many of the septic-tank systems he has designed have, with cooperation of local authorities, bent codes and ultimately led to improved code changes.

Dr. Winneberger has published extensively, chiefly on critical aspects of septic-tank systems in relation to the wide range of concerns affecting government, industry and homeowner decisions.

FOREWORD

About a decade ago, I met Dr. John Timothy Winneberger. Industrial interests then being turned towards on-site wastewater systems required association with experienced consultants in that field. Dr. Winneberger, I found, had an uncommon marriage of disciplines—natural sciences, research in sanitary engineering, and years of experience in the practical world. He is the only nationally recognized consultant I found with such credentials.

Many people simply grow older, but some gain wisdom as they mature. Friends of Dr. Winneberger who have, as I have, known him through the years, recognize him as a practical philosopher who has gained wisdom.

Federally supported researches of septic-tank systems, of which Dr. Winneberger was a major part, are classics. Two decades, from the mid-1940s to the mid-1960s, yielded research information which still is much ahead of practices today. Despite the need for better, the world of onsite wastewater management has to this day been a diffuse collection of arbitrary and capricious practices. We must not ignore scientific truths in favor of the status quo.

History testifies to the wisdom of looking to experienced practitioners for guidance where we are ignorant, the wisdom of looking to the scientific method for dispelling ignorance and revealing truths, and the wisdom of translating knowledge into practices as rapidly as possible.

About one-third of our nation's unsewered citizens depending on onsite wastewater disposal need less red tape, a recognition of true economics, a dramatic improvement in quality, and authorities with positive interests in on-site wastewater management. The need is now. Believing that there are people interested in acquiring Dr. Winneberger's yield of science and experience, friends have encouraged the presentation of his "toolkit."

It is a genuine pleasure to introduce Dr. Winneberger's works.

Fred J. Crates
Alternative Wastewater Management Consultant
Findlay, Ohio
January 1983

PROLOGUE

A formal school-text treatise on septic tanks and subsurface wastewater disposal fields has not been written, not because there are not a great many technical publications touching on the subject, but because the technology is still in its infancy. And because of the infancy of the technology, there are differing schools of thought surrounding basic issues, each with data to support its arguments. When we understand more than we do now, we will likely see where all were right in some respect, and data now in apparent contradiction will be compromised. Meanwhile, opinions stand in for knowledge, and designing a septic-tank system is either a matter of personal experience at a particular locale, or a matter of using criteria bordering on witchcraft.

FHA-SPONSORED RESEARCHES

Following World War II, there was a building boom. Thousands of new homes with septic-tank systems were insured by the Federal Housing Administration (FHA). Not all septic-tank systems functioned properly, and some people vacated their homes and defaulted on their mortgages. The FHA had to take back tracts of homes at places. Although a low failure rate, that rate constituted a major loss to the principal insurer of loans, the FHA. For the first time in history, a national agency suffered consequences of theretofore unappreciated problems with septic-tank practices.

In response, the FHA launched an investigation of septic-tank practices through the United States Public Health Service (USPHS). That investigation lasted for two decades beginning in the mid-1940s. Whereas the USPHS was assigned mostly field work, the FHA assigned to the Sanitary Engineering Research Laboratory of the University of California, the task of performing basic researches into principles of septic-tank-system function. And the Virginia Polytechnic Institute was directed to correlate septic-tank performance with soils maps, if they could be correlated.

THE AUTHOR'S PREDISPOSITIONS

The human mind is trainable, and no one is therefore free of past experiences when viewing the present. There can be no truly open mind (save a completely empty one), and it might be helpful to the reader if the writer's sources of predispositions towards septic-tank systems were displayed. A discriminating reader wants to know something of the writer whose works he reads.

Entering Research on Septic-Tank Systems

In early 1958, an unemployed plant physiologist visited Professor P. H. McGauhey, Director of the Sanitary Engineering Research Laboratory of the University of California. "Mack" asked, "What do you know about septic tanks?"

"I've lived on one, but I know nothing about them."

To my reply, Mack offered, "Well, nobody else does either. I have a small contract to study them, and I can't get an engineer interested in it. The scientific process is the same in any field and you're trained in that process. If you want the project, you can have the job."

"I'll take it, but I'll have to do a lot of reading to catch up on a subject on which I'm ignorant."

"Well," Mack mused, "A lot must be wrong with existing thinking or septic tanks would work better. There's probably an advantage to ignorance, and I would rather have a man start fresh without a lot of misconceptions."

During the first year, I gathered literature and studied it in detail. Mack was right. Misconceptions filled a vacuum of ignorance. Knowledge from other disciplines had not found its way into septic-tank technology. But most surprisingly, a device as important and deceptively complex as the septic-tank system was, had received appallingly little scientific attention. The problem was interesting.

Improving Septic-Tank Practices Became A Cause

More true than many might like to believe, government-sponsored research often suffers from over-funding and from contractual arrangements made by a modestly qualified civil servant to guide scientific inquiries of an even less qualified grantsman. If a grantsman fails to get the job, a qualified scientist is usually hamstrung by an unimaginative contract.

To my pleasure, my project was small, not overfunded such that everyone in the University wanted part of it, and I was permitted to follow my nose. Not until the final report was made was anyone much aware of the investigative courses I had chosen. Fortunately, the liaison officer of the Technical Studies Section of the Federal Housing Administration, Barney Craun, really wanted the septic-tank problem solved. And in the light of my investigative progress, Barney interpreted contractual terms as having been met. He continued to fund my studies.

As time passed, Barney noted I was continuing to make uncommon progress, worked as hard as might be expected, but I had no sense of urgency to solve the problem. I didn't know what he meant by that. Unsaid, but understood, academicians work on problems, but don't solve them. Don't misunderstand. Those scientists sometimes do good work. It's just that they only have job-scaled interests and are usually removed from the real problems. They are not on the battle front.

Back in Washington, D.C., Barney Craun arranged a field trip in hopes of impressing upon me the seriousness of the national septic-tank problem. First, I was flown to Cincinnati. At the Robert A. Taft Sanitary Engineering Center, I met scientists of the United States Public Health Service and discussed their field studies of septic-tank systems. I learned of their frustrations in trying to translate new knowledge into practical changes. Then I flew on to Washington, D.C., and met Elvin Henry and other staff of the FHA. From there, Elvin took me on a week-long tour of some of the southeastern states. We visited Professors Krebs and Hunter at Virginia Polytechnic Institute, and we exchanged information in seminar. Their efforts had been to develop techniques of using soils maps to guide septic-tank practices. They had found soils maps could successfully identify very good or very poor situations. But, unfortunately, marginal soils had not yielded to the soils-maps approach. It mattered far more what man did to such a soil during construction than what the nature of the particular soil happened to be.

Then we visited local and state authorities and learned of their efforts to cope with septic-tank problems. We conversed with a subdivider who stood to lose his locally much-needed subdivision simply because he couldn't pass percolation tests needed for FHA funding. (It wasn't the soils—it was a faulty test procedure itself.) And we viewed tracts of homes with failing septic-tank systems. Children were playing in the surfaced effluents. I was told that those homes constituted the one major life's expenditure of most of the owners. They could neither sell their homes nor afford to sewer them. They could only live with their sewages or walk away from their investments and lose them.

The point of the field trip was to impress me with the seriousness of the septic-tank problem. It was successful. Septic-tank systems were more than interesting. To about one-quarter of our nation's people who depend on them, they mattered! And later, a school teacher volunteered his summer vacation to help me in my research project. His son had drowned in a rotting septic tank.

The study of septic-tank systems became my life's work and improving practices became a very personal, almost religious cause.

From Academia To The Real World

Towards the end of almost 8 years of University-based research studies, I visited the FHA in Washington, D.C., for discussions. I found new faces and little staff knowledge of research findings to that date. My research reports were there, but nobody at hand could even find them in the building. That, and for other reasons, new research information had little affected national septic-tank practices. I was told that federal authorities could not then, for political reasons, dictate changes to local authorities—and perhaps they shouldn't. Despite several years of FHA-sponsored researches on septic-tank systems, the FHA's Minimum Property Standards remained essentially unchanged.

Most of the questions had yet to be answered, but years would pass before all was known about septic-tank systems. Knowledge was way ahead of practices, and there was a desperate need to bridge the gap. The time had come for me to leave the University, to enter private practice, and to work with real problems. There was much in the field which only an involved participant could learn.

THE TOOLKIT TO BE PRESENTED

Especially in California, there was a boom of recreational subdivisions during my first few years of practice. Subdividers soon discovered that I could pass percolation tests fair and square where the next man couldn't. There were unappreciated skills in digging holes into soils and pouring water into them. While performing thousands of percolation tests, it was sometimes possible and convenient to collect data in scientific form documenting what experience had taught me. Few of my data have been published, and I intend to present percolation-test skills from my toolkit. It would doubtless be a surprise to many technicians to learn that passing tests could have been obtained in many places where their efforts failed before. During my years of performing percolation tests at many places in our country, I have found very few soils which by using uncommonly careful methods of digging test holes and putting water into them, would not pass percolation tests (always to local surprise).

Saving distressed subdivisions by undertaking successful percolation-test programs where others had failed, provided opportunities to find applications for research findings. Both subdivider and sanitarian accepted my recommendations for improved septic-tank practices, which became required practices at the subdivision of concern. I learned that improved septic-tank practices must come from my involvement on a grassroots level, not from my prior academic position. Justifiably, local authorities suspiciously view academia and national authorities, who have no vested interests in outcomes of their recommendations. Too often, those removed people offer a mixture of practical and impractical suggestions with no way for anyone to identify the former from the latter. Suggestions which proved practical remained in my toolkit.

The FHA-sponsored investigations yielded much new information. But for the most part, concepts had not been written from a practical point of view for the field practitioner. Research reports supporting my convictions are mostly out of print. Those pieces of information, the gathering of which I was a part, are in my toolkit.

Despite our knowing that scientific truth is timeless, there is a tendency to believe that latest information must be best. Nevertheless, earlier investigations have often been better performed than more recent repetitions. Thus I offer what I believe to be best data, without respect to the date of particular researches.

Presentations are not to be an accounting of all points of view. Man is a subjective creature. No writer involved with his topic could present contradictory views as fairly as the other fellow could. It is my intention to convey to the reader, septic-tank practices from my point of view. Corroborative information will be presented where believed helpful. And, of course, information will be offered from selected researches of others, which is to round out the general topic at hand.

STYLING

Early scientists often wrote in relaxed styles. Discriminating readers could easily distinguish factual data from a writer's personal view of matters because there was no pretense of the early writer to be other than a subjective creature.

Deplorably, I believe, there has been a development of scientific writing styles in which writers simulate objective machines. They are not machines, of course, and their dispositions are unconsciously and subtly woven into the fabric of both their investigations and their writings. Readers often have a difficult time in sorting matters, and the layman is often misled. After all, craftsmen, including scientists, come in qualities, and it is not helpful to have work of questionable quality disguised by simulated objectivity.

For the foregoing reasons and because readers include a wide variety of disciplines, each with its own jargon, I prefer to emulate the relaxed style of early scientists. But, the reader is encouraged to bear in mind that presentations combine information derived from scientific inquiries with opinions derived from experience. Each of those kinds of information will likely be easily distinguishable, one from the other.

"GOING METRIC"

In feeble efforts to change from the English system of measurements to metric, authors frequently include metric units in parentheses after their English units. The results are often preposterous metric units; for example—3 gal./sq.ft./day (11.36 *l*/0.0929 sq. meter/day). The affect on metric readers is equal to, say, Fritz "goes English." Here's his advice on a fertilizer coverage—Benutzen 1/2 kg/m^2 (1.34 lb./1.2 sq.yd.).

In any country, formulas, manufactured nuts and bolts, recipes, etc., are developed in convenient terms of the system of units. English and metric systems are not alike. So a convenient number in one system generally translates into a weird fraction in the other system, which no one could conveniently use. And, not uncommonly, someone will translate into unfamiliar terms which are ludicrous to the person familiar with that particular system. How about an endurance snail race won in 1.98 furlongs/fortnight?

Therefore metrics are not included in text. The writer knows metric units and doesn't need the exercise in translating English units to metric, readers know English units and can convert them to suit themselves, and domestic readers find impertinent metrics in parentheses distracting.

John H. Timothy Winneberger
Santa Fe, New Mexico

CHAPTER 1

ANATOMY OF A SEPTIC-TANK SYSTEM—TERMINOLOGY

WASTEWATER GENERATION

Household sewages, nowadays termed *wastewaters*, include everything that moves down the house drain. Because of recent interests in waterless toiletry, liquid and solid body wastes have become known as *black wastes* (black being a symbol of evil), and those wastes together with water used for flushing become *black waters*. *Grey waters* must be defined in terms of what they are not, that is, all household wastewaters less toilet wastes. Grey waters can be relatively clear and innocuous. Few health authorities are moved to action when viewing laundry waters bypassing household plumbing to irrigate the backyard. On the other hand, some cooking habits result in greywaters stronger than many ordinary sewages in grease contents.

Dry toilets and grey water treatment systems can serve properly. But in the practical world, there is a profound difference between the questions, "Can it work?" and "Will it work?" Even advocates of dry toilets and grey water treatment arrangements tire of maintenance. Thus, a study conducted by the California State Department of Health Services discovered most such systems fail to meet performance criteria [1,2].

Septic-tank systems are the topic at hand; therefore all household wastewaters combined are the assumed influents to the systems. There are available treatises on wastewater characteristics and flow volumes to which the reader is referred [3–6]. The designer of septic-tank systems will find many local authorities mandating estimates of expected wastewater volumes. If not, the reader would best make efforts to learn what his particular situation is likely to be. Wastewater volumes vary far too widely for valid use of common values.

BUILDING SEWERS—CONDUCTORS FOR DISEASE VECTORS

The *building drain* receives waste discharges collected by other drainage pipes of the building and conveys them to the *building sewer*, which begins 3 ft (or as locally defined) outside of the building foundation. Ordinarily the building sewer is a solid pipe with a diameter of 4 in., and it is laid to a downgrade of 1/4 in./ft.

The entire system of drains contains air as well as occasional wastewater flows. Whereas a few decades ago a building sewer had a water trap just outside of the house [7], today's building sewer has no water trap separating the atmosphere of the building sewer from the atmosphere outside of the open, unscreened vents on the roof. It is known that disease vectors (flies, etc.) can and do live within septic tanks. When ready, those disease vectors find an uninterrupted pathway to the outside world through the building sewer and out of the house vent. That situation—with an easy remedy—is discussed in Volume II. Health authorities in Australia have been concerned with sewage-oriented disease vectors being cultured within human habitations. Those authorities require vents to be screened such as to prevent free traffic of disease vectors between human habitations and sewage environments [8]. American health authorities would do well to mandate likewise.

An extensive market survey was performed by one of the writer's clients a few years ago. It disclosed that odors (or the potential for them) constituted the first, overwhelming concern of homeowners with their septic-tank systems. Fear of improper operations was a mild second concern. Odors, caused primarily by hydrogen sulfide, are gaseous. Odors, like disease vectors, are ordinarily given an uninterrupted pathway from the septic tank to the house vent. A few authorities require capped tees or ells over inlet devices to confine odors to the septic tanks.

By the bye, experience has been that waste clogging of building sewers is less likely to occur in straight or gently curved sections of pipe. It is the sharp bends that generally cause troubles.

SEPTIC TANK

The building sewer provides wastewater flows *influent* to the *septic tank*. That tank is a sometimes watertight, covered receptacle designed and constructed to:

- receive wastewaters from the building sewer.
- separate settleable *sludge* and floatable *scum* solids from the through-passing wastewaters.

- store or digest those separated substances.
- pass treated wastewaters out of the tank as *effluents*.

As will be elaborated in Volume II, sludges are black and viscous fluids. They are composed of particles which are negatively charged when comparatively freshly disposed to the tank. As sludge particles age for some years, they seem to become positively charged. At least they tend to respond better to anionic rather than cationic flocculants [9,10].

Periodically, sludge and scum must be removed from septic tanks. That is the function of septic-tank pumpers (termed pumper because of the use of vacuum pumps). Although different authorities advise cleaning (pumping out a septic tank) on yearly to perhaps 7-yr schedules, measurements of accumulations in tanks managed by public entities in Florida and Oregon indicated about a decade or so may pass between needed cleanings of tanks on the order of 1000 gal liquid-holding capacities [11,12].

Where proper maintenance of the septic tank is planned, the lids to the tank should not be buried with the rest of the tank. Rather, *risers* (or manholes) should be provided, each of which may consist of a concrete block, wooden, or plastic passageway extending from the lid of the tank to above ground surface. Risers should be fitted with lids removable by a serviceman, but not easily removable by children or other unauthorized persons. Odors can be kept within the tank by sealing the lids with silicone. Those sealed lids are easily opened and again resealed.

As said, septic tanks are sometimes watertight. They are always intended to be so. Poor construction, breakage, and the like, render many tanks leaky. Also, it is common knowledge that contractors find it easier to punch a hole in the bottom of a tank floating in high ground waters than to sink the intact tank by waterfilling it.

A leaky septic tank can lead to troubles. During a household vacation or other period of cessation of wastewater flows, a leaky septic tank may drain to below the outlet device. Then upon resumption of influent flows, the rising scum mat can enter the bottom of the outlet device and pass out with effluents to clog the system [13].

A second cause of troubles with septic tanks comes from their great weight, together with customary poor installation practices. A filled 1200-gal tank weighs 5 tons plus the weight of the concrete tank itself. Rather than set them on a proper bedding of gravel, sand, or concrete slab, contractors generally place empty septic tanks on the loose soil at the bottom of the tank excavation. Water filling the tank usually waits for house occupancy. Then, commonly, the tank settles and may shear off inlet and/or outlet pipes. Authorities requiring a water-filled tank for testing watertightness, in that way reduce chances of shearing pipes from settling.

CESSPOOL—A KIND OF SEPTIC-TANK SYSTEM

Generally when one speaks of a *septic-tank system*, he refers to a septic tank followed by an arrangement for subsurface disposal of effluents into soils. Because septic-tank systems are so generally used to manage household wastewaters, the expression sometimes is misused to include other devices as well.

A *cesspool* is an excavation in a deep soil, generally a dozen feet, or so, deep. Ordinarily, it is cylindrical with several feet in diameter. There is a porous inner wall which may consist of masonry, precast concrete rings, or some other material offering strength for shoring the excavation. Often the annular space between the inner wall and the outer soil wall is gravel filled. The cesspool ordinarily has a masonry dome over which soil is backfilled to grade. The building sewer terminates at the top side of the inner wall wherein raw wastewaters flow influent to the cesspool. It is intended that wastewaters seep away through the surrounding soils and solid debris be stored or digested within the inner wall.

Only in geometry does the cesspool differ from the septic-tank system with its separate, gravel-filled, subsurface disposal field. Functionally, the inner wall of the cesspool is a leaky septic tank, and the outer soil wall is the subsurface disposal field. People interested solely in construction details will regard cesspools and septic-tank systems as different devices. People interested in functions will see few differences.

Nevertheless, a cesspool is limited in that treatment and disposal are intimately connected. Geometry of excavations is limited, and there is no practical way to spread disposal of effluents over a comparatively wide area such as a separate disposal-trench system can do. In that respect, rather than because of some functional difference, the septic-tank system may be advantageous. Most codes forbidding cesspools exist without technical justifications.

INJECTION SYSTEM—THE DISPOSAL FIELD

The purpose of the man-made injection system is injection of septic-tank effluents into the natural soil system of a site. This is accomplished by the *disposal field*, which customarily consists of a hole in the ground of some regular geometry, has a backfilling of gravel or stones to maintain the integrity of the hole, and has a system of pipes believed to distribute wastewaters throughout the hole in the ground. It is this injection system which becomes clogged in time and has received most of our design attention. Maximum wastewater loading rates have been related to the injection system, the effluent-soil interface. Being somewhat clogged from wastewaters, the injection system is much oversized compared to what would have been needed for short-term acceptance of fresh waters.

Gravels (drain rock, etc.) used to backfill disposal fields are ordinarily quite coarse. When smaller gravels are recommended to an authority, he sometimes believes less storage space for ponded effluents would result. The pore size will be smaller, but the total *porosity*, the proportion of void to solid volumes, is not a matter of gravel size (see Figure 1). Two same-sized piles of perfect spheres, say one of cannon balls and the other of buckshot, will have exactly the same porosity.

Gravels will vary in porosity because of their irregular shapes and gradings. Crushed and sorted gravels have porosities on the order of about 43%. When porosity must be known, it can be determined by pouring measured amounts of water into a known volume of gravel. Use a kitchen measuring cup with a bucket of known dimensions full of gravel.

When the disposal field element is about rectangular in cross section, quite long, and not exceptionally deep, that element is termed a *disposal trench.* The disposal trench contains a gravel filling over which is placed a *barrier* of straw, building paper, or other material to exclude soil used to backfill the trench to ground level.

Within the gravel filling, distribution pipes of past years consisted of short lengths of clay drain tiles (or other material), loosely butted

Figure 1. A pile of sand grains or gravel? The drawing could represent either. The porosity—the proportion of void to solid space—would be the same.

together. Hence the early term *tile fields*. In recent years, perforated plastic pipes commonly serve in disposal trenches (also termed leaching field, drain field, absorption field, soil disposal field, etc.).

When the width of a disposal trench is greater than about 3 or 4 ft, a *disposal bed* (seepage bed, etc.) is had. The disposal bed originated because of the ease with which it could be dug with a bulldozer or perhaps a front-end loader. Like disposal trenches, beds have barriers separating gravel below from soil above. But whereas the trench contains one distribution pipe, the bed may contain several pipes in satisfaction of local requirements of minimum spacing of those pipes.

When the geometry is a vertical cylinder, it is termed a *disposal pit*. Occasionally a very deep trench is also termed a pit. To maintain the integrity of the excavation, a disposal pit may contain precast concrete rings, a cylindrical masonry wall, gravel, or larger stones.

Old disposal pits have caused needless loss of life which could have been obviated by safety-conscious authorities. All pits should be gravel filled around a distribution pipe and under a barrier, and then soil filled to ground surface.

Whereas authorities generally require a distribution pipe running the length of a trench, they often do not require the pipe in a vertically oriented, gravel filling of a pit. A perforated distribution pipe should run from top to bottom, within a gravel-filled pit. Clogging of the gravel surface receiving effluents has been observed where perforated pipe within gravel had been omitted [14].

Breezers and Observation Pipes

Apparently originating in Canada, sometimes the perforated distribution pipe is fitted with vertical sections of solid pipe rising above ground level. Those *breezers* (or *breathers*) are fitted with protective caps, but ones through which air can freely flow.

A straight, solid pipe extending aboveground and perforated within the gravel-filled excavation to its bottom, and fitted with either a removable, vented or solid cap, is an *observation pipe*. Those pipes are used where observations of liquid levels within the disposal field are of interest.

Infiltration Surfaces

Soil surfaces in a disposal field which can be inundated with effluents are the *infiltration surfaces* (also active absorption areas, area requirements, etc.). But because of the historical way effluent-loading rates have been related to disposal-field design, authoritative mandates often designate bottom surfaces of disposal trenches or sidewalls of pits as infiltration surfaces to be credited. The U.S. Public Health Service (USPHS) *Manual of Septic-Tank Practice* [15] equivocates by relating effluent loading rates to bottom areas of trenches, while at the same time claiming, "Absorption area is figured as trench bottom area, and includes a statistical allowance for vertical sidewall area." It does not.

The U.S.EPA *Design Manual, Onsite Wastewater Treatment and Disposal Systems* [16] asserts the designer may credit both bottoms and sidewalls of disposal fields. Then the authors equivocate with unconvincing arguments as to when which serves. Incidentally, authors of that EPA document state, ". . . the clogging mat can slough off the sidewall." It cannot. The complexities of effectiveness of sidewalls vs bottom surfaces are not predicated on such a naive concept. Perhaps the EPA authors assimilated a similarly erroneous statement of McGauhey [17,18] to the same effect. (Because of heavy use of his researches, the writer was courteously credited as coauthor of those particular documents actually written by McGauhey.)

Although excessively settled septic tanks tolerate liquid levels higher than inlet and outlet plumbing devices, it is best to regard the intended liquid levels in tanks in hydraulic-head considerations. Then, any soil surfaces of a disposal field which are equal to or less high than the outlet device could serve as infiltration surfaces. Nevertheless, many local authorities would find it impractical to have requirements based on such a thought, however simple it may seem. Those authorities generally would prefer to rely on distribution pipes being gravity-served, with effluents and soil surfaces below the distribution pipe in the disposal field to be credited in satisfaction of infiltrative-surface requirements.

DISTRIBUTION

Effluents from the septic tank enter a solid *distribution pipe* (also termed distribution line), 3 or 4 in. in diameter. It conducts effluents to a distribution box, or to a wye or tee, or straight away to the subsurface disposal field. Within the field, distribution pipes are perforated nowadays, but in past years pipes within the field consisted of lengths of vitreous clay (or concrete) tiles loosely butted together. That entire system of pipes is the *distribution system.*

Distribution Box

The *distribution box* is a comparatively small box with a lid, which serves as a manifold. The pipe from the septic tank conducts effluents into the box. Pipes from the distribution box leading to various segments of the disposal field are supposed to receive equally divided portions of effluents from the box.

Theoretical and practical studies disclosed that distribution boxes cannot and do not equally divide effluent flows [15,19]. Rather, they preferentially and capriciously guide flows to one pipe now, and to another later. A leading USPHS investigator commented, "Other calculations as well as several thousand observations lead us to the conclusion that anything approaching perfect distribution is a practical impossibility with the distribution box," [20]. When the distribution box leads to a disposal field in flat terrain, uneven distribution is of no consequence. But in sloping terrain, the box can lead to a premature failure [21]. Effluents preferentially flowing downslope to one element of the disposal field soon overload that element, and then it surcharges. When effluents cannot back up the pipe to the box without causing ponded effluents to breach the soil covering of the element, effluents reach the ground surface and cause unacceptable dampness and odors.

The distribution box has a usefulness. It provides a central point at which to exercise some control over the system. Repair elements of the disposal field can have their distribution pipes more easily inserted into knockouts in the box than joined to existing pipes.

Parallel Distribution

The intention of *parallel distribution* is the serving of all elements of a disposal field with effluents equally divided among them.

The distribution box is the device intended to accomplish parallel distribution. In level terrain, the distribution box, or the use of pipe fittings such as wyes and tees, accomplishes parallel distribution after a disposal field has aged such as to contain a more or less permanent pond. Although effluents may flow preferentially down one pipe, effluents can easily back up to flow down another pipe.

Serial Distribution

The intention of *serial distribution* (also termed step down, land block, and others) is the directing of all effluents to first one element of a disposal field. When that element fills, surplus effluents then go to the next element in turn, and so on. Serial distribution is accomplished by an overflow pipe from one trench to the next below (Figure 2), or by a drop box.

A *drop box* is a small box outside of the disposal-field element. It has

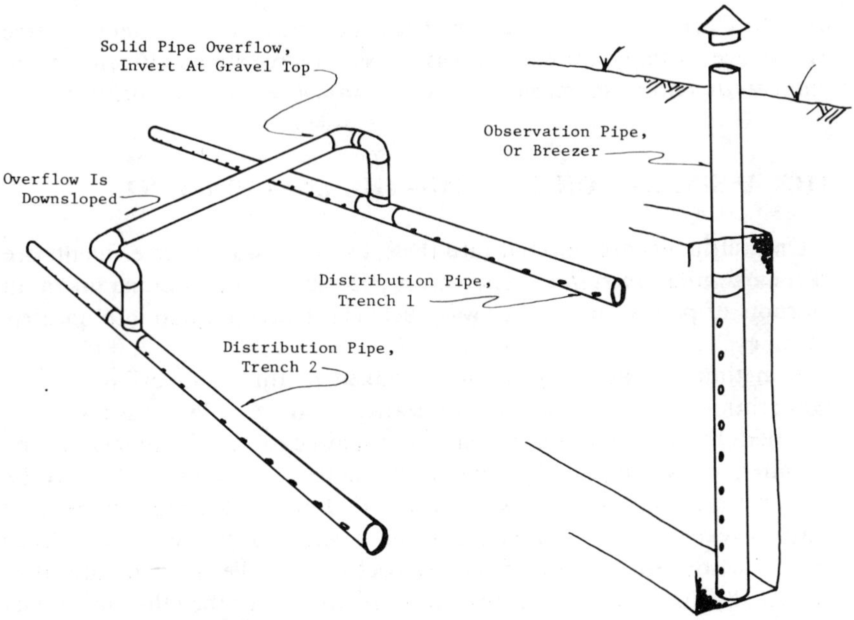

Figure 2. Serial distribution by an overflow pipe, and an observation pipe or breezer.

one pipe leading to that element. At a higher elevation in the drop box, another pipe leads to another element. Flows influent to the drop box preferentially go to one element. When it is filled as intended, flows eventually back up into the drop box, rise in elevation, and then they flow down another pipe to the next element below. Generally, all elements of the disposal field have one drop box each. As in the case of distribution boxes, drop boxes provide points of entry to the system.

Pressure Sewer

Occasionally, gravity flow to a disposal field is impossible. There a sump is fitted with a pump. Through a *pressure sewer*, ordinarily a plastic pipe of relatively small diameter (usually matching the pump outlet), effluents from the sump are conveyed to the disposal field. Arriving at the field, the pressure sewer generally discharges into ordinary distribution pipes.

Perhaps someone has neglected cleaning his septic tank and solid debris accumulations have carried over and clogged his pump and pressure sewer. Nevertheless, the writer has never heard of anyone having need to clean his pressure sewer. That conforms with maintenance-free experience with community pressure sewers receiving effluents from *interceptor tanks* (so termed are septic tanks in such applications).

THE ASSIMILATION SYSTEM—THE ENVIRONMENT

Once effluents are injected into the soils of the site, those effluents are in the *assimilation system*. They must percolate away, become taken up by roots of plants, or in other ways become reincorporated into nature's water cycle.

Sometimes *evapotranspiration* is looked to for assimilation, that is, evaporation from soil surfaces or transpiration by plant tissues.

When the assimilation system is impaired or is threatened to be impaired by a temporarily high water table, a *curtain drain* may be constructed. That drain consists of a buried perforated pipe, sometimes gravel packed, intended to collect ground waters or intercept them when they flow towards the site of the disposal field. Collected ground waters are usually discharged to nearby drainage ditches, or they flow by gravity to daylight when on hillsides.

CHAPTER 2

PERCOLATION TEST

The percolation test, whatever the local variety may be, is far from being a reproducible standard. Skills of technicians, often more than soils themselves, affect outcomes. There is a desperate need for a valid percolation test which will yield the same result for one technician as for another. The test exists. Methods exist. The answer is return to Henry Ryon's original test and promulgation of proper field methods.

In 1962, during the FHA-hosted field trip mentioned in the prologue, Cousin Bill was visited in Virginia. He was a local developer of a much needed, but distressed subdivision. To persons acquainted with soils and successful septic-tank practices nearby, Cousin Bill's soils seemed obviously sutiable. But, his "perc" tests had not met FHA standards, and consequent financing difficulties were in sight. Being with the University of California and located on our nation's opposite shore, the writer and his colleague could only point out shortcomings of misleading percolation tests [21]. The writer and his colleague were not in a position to direct needed work.

Two decades later, Eleanor arrived in California from Alaska. She paid $35,000 in cash for a lot on which to build her dream house. When the lot failed to pass its percolation tests, Eleanor faced serious loss. This time the writer was a help. Through development of field techniques for getting best percolation-test results of which soils were capable, the writer had become known as the man to call when "perc tests" had failed. "He can get your lot or subdivision to pass percs, fair and square." Eleanor's lot passed.

Over years of performing thousands of percolation tests, there were opportunities to collect data which no government-funded project had undertaken to collect, so the information is not in the literature. The time has come to share knowledge with others.

If a technician wants to make a living by becoming known as the man to get a lot or subdivision to pass percolation tests where the next technician may fail, read carefully on.

If an authority wants an easily performed, conservative percolation test, defendable if ever taken to task, read carefully on.

RYON'S ORIGINAL PERCOLATION TEST

A misnomer, the percolation test measures neither percolation nor infiltration, *percolation* being movement of water through a soil mass and *infiltration* being movement of water into a soil surface. The test is an empirical procedure for observing water seeping into a hole in the ground, and it has been empirically related to loading rates (criterion of design size) of septic-tank effluents onto soils of a subsurface disposal field.

Henry Ryon devised the original percolation test when he was affiliated with the former New York State Engineer's Office, and then continued while he was Senior Sanitary Engineer with the New York State Department of Public Works. Ryon wrote [22]:

> Soil Test—A practical but rather empirical test to determine the absorptive quality of the soil which has proved very satisfactory may be made as follows: A hole about 1 foot square and 18 inches deep (the depth of the proposed trenches) is dug at the site of the proposed field. This hole is filled to a depth of about 6 inches with water, taking care to wet the soil before pouring in the water for the test if it appears dry. (It is well to repeat the test to eliminate the effect of a very dry soil.) The time that it takes the surface of the water in the hole to fall one inch is then observed.

The word empiric is from Latin and Greek, which is unfortunate. Anglo-Saxon words have the clearest meanings in the mind of an English-speaking person. Empiric is neither well understood as a word nor as a concept. As for the empirical yardstick, change the length of the inches and one has a yardstick which he can say is one yard long, but his yard will be different from another person's yard. As for Ryon's empirical test, change his procedures and one will have a percolation rate which he may report to authorities as so many minutes per inch (min/in.), but his min/in. will be different from the min/in. Ryon would have measured.

Ryon described his test in simple terms. One would gather that Ryon used a shovel rather than a power auger to dig the hole. He was likely simulating water in a 1-ft-wide disposal trench, those he studied, and he made no effort to simulate the soil-destructive forces of heavy machinery often used to construct disposal fields. Ryon likely did not keep adding presoaking waters into the hole for several hours, commonly prescribed nowadays; rather, he likely filled the hole once in a slowly permeable soil, or perhaps twice in a faster soil. Especially note that Ryon permitted a physical occurrence, the infiltration of the 6th inch of water over a 1-ft-square hole bottom, to be a constant happening, while he measured time as a variable.

THE USPHS PERCOLATION TEST

During the mid-1940s, the United States Public Health Service (USPHS) entered a 2-decade-long series of field studies of septic-tank practices. One effort was to confirm the usefulness of Henry Ryon's percolation-test criterion, that is, the empirical calibration of the test to the design size of a disposal field. Those investigators concluded Ryon's criterion was valid. But when the USPHS *Manual of Septic-Tank Practice* was produced [15], authors unwittingly changed Ryon's test to a more liberal and capricious one. The USPHS percolation test is as follows:

> 1. *Number and location of tests.*—Six or more tests shall be made in separate test holes spaced uniformly over the proposed absorption field site.
> 2. *Type of test hole.*—Dig or bore a hole, with horizontal dimensions of from 4 to 12 inches and vertical sides to the depth of the proposed absorption trench. In order to save time, labor, and volume of water required per test, the holes can be bored with a 4 inch auger.
> 3. *Preparation of test hole.*—Carefully scratch the bottom and sides of the hole with a knife blade or sharp-pointed instrument, in order to remove any smeared soil surfaces and to provide a natural soil interface into which water may percolate. Remove all loose material from the hole. Add 2 inches of coarse sand or fine gravel to protect the bottom from scouring and sediment.
> 4. *Saturation and swelling of the soil.*—It is important to distinguish between saturation and swelling. Saturation means that the void spaces between soil particles are full of water. This can be accomplished in a short period of time. Swelling is caused by intrusion of water into the individual soil particle. This is a slow process, especially in claytype soil, and is the reason for requiring a prolonged soaking period.
>
> In the conduct of the test, carefully fill the hole with clear water to a minimum depth of 12 inches over the gravel. In most soils, it is necessary to refill the hole by supplying a surplus reservoir of water, possibly by means of an automatic syphon, to keep water in the hole for at least 4 hours and preferably overnight. Determine the percolation rate 24 hours after water is first added to the hole. This procedure is to insure that the soil is given ample opportunity to swell and to approach the condition it will be in during the wettest season of the year. Thus, the test will give comparable results in the same soil, whether made in a dry or in a wet season. In sandy soils containing little or no clay, the swelling procedure is not essential, and the test may be

made as described under item 5C, after the water from one filling of the hole has completely seeped away.

5. *Percolation-rate measurment.*—With the exception of sandy soils, percolation-rate measurments shall be made on the day following the procedure described under item 4, above.

A. If water remains in the test hole after the overnight swelling period, adjust the depth to approximately 6 inches over the gravel. From a fixed reference point, measure the drop in water level over a 30 minute period. This drop is used to calculate the percolation rate.

B. If no water remains in the hole after the overnight swelling period, add clear water to bring the depth of water in the hole to approximately 6 inches over the gravel. From a fixed reference point, measure the drop in water level at approximately 30 minute intervals for 4 hours, refilling 6 inches over the gravel as necessary. The drop that occurs during the final 30 minute period is used to calculate the percolation rate. The drops during prior periods provide information for possible modification of the procedure to suit local circumstances.

C. In sandy soils (or other soils in which the first 6 inches of water seeps away in less than 30 minutes, after the overnight swelling period), the time interval between measurements shall be taken as 10 minutes and the test run for one hour. The drop that occurs during the final 10 minutes is used to calculate the percolation rate.

The USPHS test differed from Ryon's in specifying a deeper water filling to start; that is, 6 in. of water over 2 in. of gravel is 8 in. of water over the actual soil bottom of the hole. The USPHS test specified a round hole from 4 in. to 12 in. in diameter whereas Ryon's 12-in.-square hole would have been about equivalent to a round hole 13.5 in. in diameter.

The USPHS offered a purpose to presoaking which did not likely differ from Ryon's thoughts. A 4-hr period of observations was directed. But where it was intended to have water refilled to the mark each half hour, ambiguous wording, ". . . refilling 6 inches over the gravel as necessary," has been misinterpreted by some authorities as meaning to refill only if the hole threatens to run dry (an unneeded caprice to hydraulic head).

A profound source of caprice to the USPHS test was holding time as a constant, the last half hour of a 4-hr period, while assigning physical change, water seepage, to a variable status. Water receding in a hole is a decelerating movement. The parallel would be a horse race: whereas Ryon would measure time needed to run a given distance, the USPHS would run the horses a given time and then see how far they went.

OTHER PERCOLATION TESTS

Throughout California, perhaps more so than elsewhere, there has existed a large variety of percolation tests. Some authorities have been openly aware of their local test being different from the USPHS test. Other authorities have directed the USPHS test in one paragraph of a handout, but then gone on to describe contradictory methods.

In the San Francisco Bay Area are five counties. The depths of waters over the test-hole bottoms in those counties have been, 8 in., 8 in., 12 in., 14 in., and about 2 ft [23]. There are other major differences in procedures as well. Regional authorities recognizing the confusion have recommended a return to Ryon's original test [24].

Ludwig's Modification

Ludwig and co-workers [25,26] directed examination of soils to select the most permeable. Then a hole 1 ft square is to be constructed 18 in. deep into the stratum of interest.

> The test is made by filling the hole with approximately 12 in. of water, the exact distance to the level of water being noted and also the time of this initial filling. Depending upon the rate at which the water level appears to be dropping, a period of time, designated as (t), such as 30 minutes or 60 minutes, is selected as the interval at which repeat measurements are to be made. After elapse of this selected period of time, the distance to water level is measured, the hole then immediately refilled to approximately its original 12 in. depths, and the distance to water level again measured. This procedure is repeated until the rate of drop appears to be leveling off. A total period of testing of not less than 6 hours is recommended.

Ludwig and his students had found that as time passes, successive water drops become more the same. They approach a stabilized rate. When graphed, such data form a curve approximating a hyperbola, and mathematical treatment can produce a graph with the hyperbola straightened out to a straight line. The slope of that line is the asymptote, the equilibrium rate one should measure theoretically, if Ludwig's test were performed indefinitely. Ludwig presented a mathematical treatment for a graphical solution to finding the equilibrium percolation rate.

In simpler arithmetic than Ludwig offered, an example of management of data is presented in Table I.

In Figure 3, data from Table I were graphed. Then a straight edge was used to draw a line believed best to fit the data points. The slope of that line was then calculated by picking two convenient points towards opposite ends and reading the values on the T axis for numerator and T/R axis for the denominator of the given fraction. The quotient, 4.1 min/in., is the equilibrium rate.

Notice in Table I that the last four readings were alike, d = 4.0 in. Those water drops were at R = 3.75 min/in. One would likely have continued to

Table I. Lizard Estates, Test Hole 12, Worksheet for Ludwig's Modification[a]

t	h_1	h_2	d	T	T/R
15	12.0	4.4	7.6	7.5	3.8
30	12.0	6.7	5.3	22.5	8.0
45	12.0	7.5	4.5	37.5	11.2
60	12.0	7.8	4.2	52.5	14.7
75	12.0	8.0	4.0	67.5	18.0
90	12.0	8.0	4.0	82.5	22.0
105	12.0	8.0	4.0	97.5	26.0
120	12.0	8.0	4.0	117.5	31.3

[a]t = time in minutes from the start; h_1 = initial reading of a float gauge; h_2 = final gauge reading at the end of the selected time interval; d = water drop in inches (h_2–h_1); T = time in minutes from the start to the middle of each successive time interval (ignoring time required for water refilling); R = rate of water drop, min/in.; and I = time of the interval in minutes selected for successive water drops. Note that values of T were figured to the middles of t values. That is: T = 7.5 is the middle of t = 15; T = 22.5 is to the middle of t = 15 and 30; etc. T/R values were calculated by: T/R = d × T ÷ I (in the example I = 15 min.).

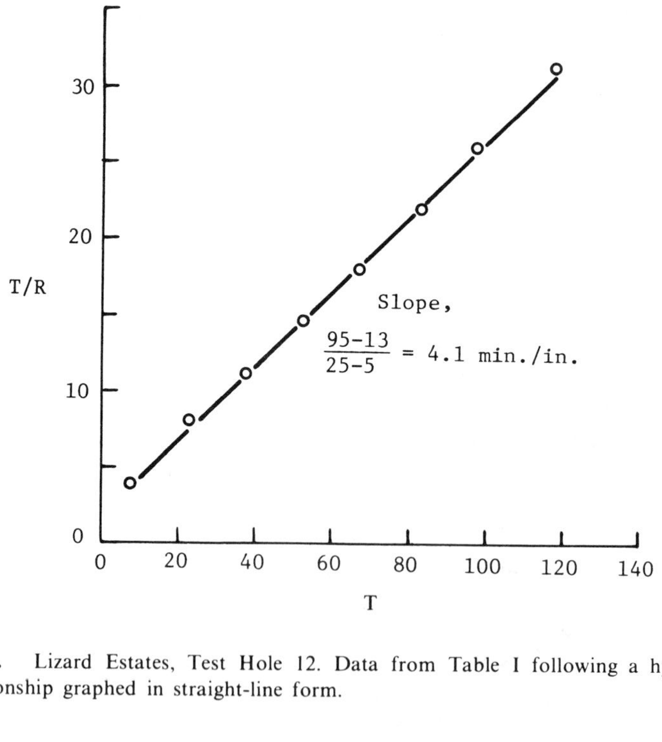

Figure 3. Lizard Estates, Test Hole 12. Data from Table I following a hyperbolic relationship graphed in straight-line form.

test a long time before the graphically measured equilibrium rate of 4.1 min/in. would have been actually obtained in the field.

In viewing Figure 3, one may notice a tendency for the data to delineate a slightly waving line, rather than an absolutely straight one. That tendency has been a common finding.

In this writer's experience, probably because of his careful techniques, field measurements for Ludwig's Modification always required only an hour or two, and the time was always shorter than the USPHS test would have required. Still, the equilibrium rate was always only slightly slower than the last measured rates after a test stabilized. And, for various reasons—the 12 in. of water fillings for one—Ludwig's Modification always yielded a faster percolation rate than either Ryon's test or the USPHS test would have yielded.

Ludwig's Modification will yield a different equilibrium rate for different choices of a time interval for successive water drops. That is, Ludwig's Modification suffers the same deficiency the USPHS test suffers. It sets time as a constant, while permitting physical changes to vary.

A Refinement

Mathematics can be used to measure equilibrium percolation rates of tests performed with time a variable and physical change a constant.

If successive drops of water at the same increment of depth in a hole are timed, those successive time periods will approximate a hyperbola when graphed. Mathematics can then produce a straight-line graph and the slope of that line will be the asymptote of the hyperbola. That is, the slope will be the time period observed after the selected water drop has been observed an indefinite number of successive times, assuming soil permeability and other matters remain constant.

During studies of a subdivision in California, Ryon's percolation test was performed with repeated measurements. Table II displays the field data prepared for mathematical analysis. Because a 1-in. water drop was successively timed, the recorded time in minutes is also R, the rate in min/in. The cumulative time to the middle of the successive time interval T was calculated by dividing the rate by 2 and adding the quotient to the cumulating values of T. That omitted time needed for water refillings from considerations.

The data in Table II, when graphed in Figure 4, display a hyperbola. The asymptote (to be solved for) has been drawn on the graph. The asymptote is the value which would theoretically be reached if an infinite number of data were collected.

Table II. Rocky Estates, Test Hole 33, Worksheet for a Refinement of Ryon's Percolation Test

R	T	T/R
5.75	2.88	0.501
18.50	12.13	0.656
27.17	25.71	0.946
32.50	41.96	1.291
35.75	59.84	1.674
37.75	78.71	2.085

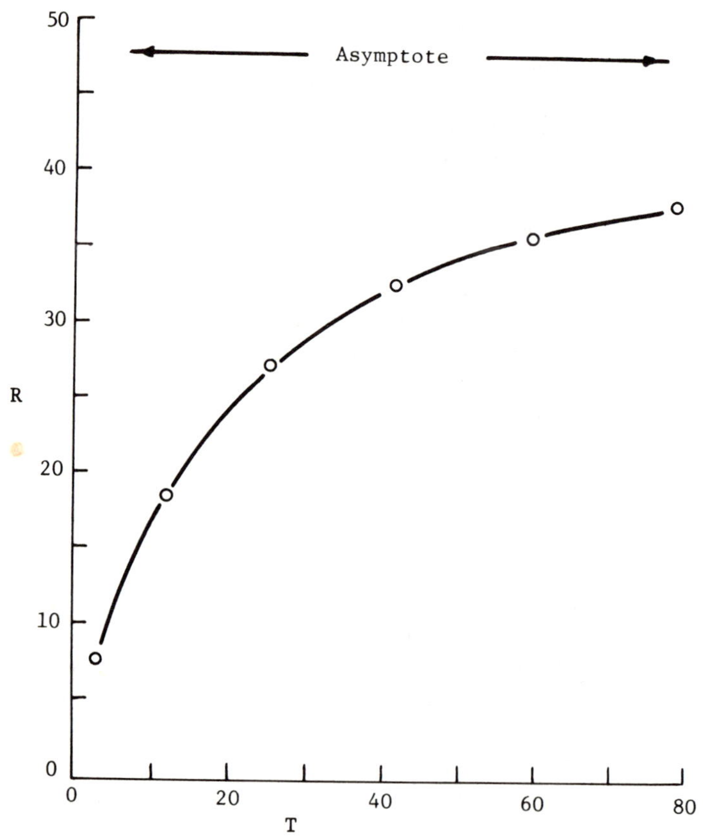

Figure 4. Rocky Estates, Test Hole 33. Data from Table II display a hyperbola. The asymptote is the value the hyperbolic curve reaches at infinity, i.e., an infinite amount of data collected.

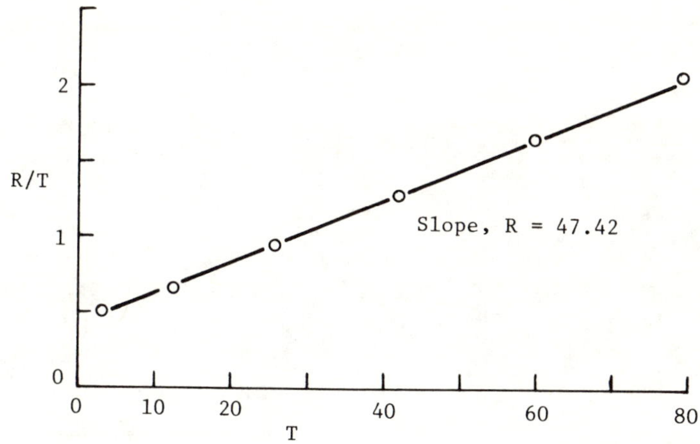

Figure 5. Rocky Estates, Test Hole 33. Data from Table II following a hyperbolic relationship graphed in straight-line form.

Figure 5 displays the data in straight-line form. Whereas the slope of the line was found graphically in Figure 3, the slope of the line in Figure 5 was found by statistics of regression. With a hand-held computer, the data in Column R of Table II can be entered and the equilibrium rate computed, without resorting to the time-consuming graphical solution.

The reader is notified that the writer does not advise routine field use of this refinement or Ludwig's Modification of the percolation test. Both amount to far more efforts towards accuracy than either variations in test results or anticipated use of data would justify. Use of this refinement or Ludwig's Modification would be equal to estimating the soil area needed in a disposal field to a fraction of a square inch. These arithmetical approaches to the percolation test were presented only out of academic interests.

CHAPTER 3

TECHNIQUES FOR BEST PERCOLATION TEST RESULTS AND ASSESSMENT OF SOIL PERMEABILITY

Always remove the upper soil layers of no concern to expose the soil to be tested. The test hole is to be at the bottom of a hole large enough for a technician to work within. That is most easily accomplished with a backhoe. The percolation test hole is to be hand dug to about a couple of inches deeper than water fillings of the test are to be.

DIGGING THE TEST HOLE

At the bottom of the larger hole, dig a test hole smaller in diameter than the finished test hole is to be. A bucket auger 3¼ in. in diameter is manufactured for soils work [27], and it serves quite well for testholes that are to be 4 in. or so in diameter. A posthole digger will serve if the hole is to be 12 in. in diameter. Take care when digging the smaller hole not to crush the soil sidewalls.

With a 2-in. rigid putty knife in hand, insert the blade into the top side of the hole opposite you, holding the blade with its cutting edge vertical. Pull the blade away such as to break loose a chunk of soil. Next, insert the vertical blade into the soil at a place perhaps an inch away from the place where the chunk was just removed. Break loose another chunk of soil. Continue working around the hole until back at the start. Then the blade is used likewise to remove a next ring of soil below.

As one works down into the hole, soil having fallen to the bottom must be removed carefully by hand or with a small dipper, taking care not to brush against the freshly exposed sidewalls of the hole.

Depending on the size of the hole, several wider rings may need to be removed before the finished diameter is had. The bottom should be almost flat. Also, select the smallest hole diameter local practices permit.

The objective in hand-digging the test hole is to have a hole with sidewall soils in as nearly an undisturbed state as possible. Breaking soil away best meets that objective. Never cut the soil in a shearing, chisel-like manner. Ideally, most of the hole sidewall will not have had contact with the putty knife, and no crushing forces will have been suffered by the soil. The bottom of the hole needs less care.

When the hole is finished, it will have ragged, irregular sidewalls of soils with pores and small roots (if present) in view. It will look much like a clod of soil will look if one breaks the clod and exposes a fresh soil face.

PAPER BASKET LINER

Cut the bottom out of a large paper bag and cut the bag open along a side. Lay the opened paper bag onto grass, onto a soft garment, or otherwise support the paper such that a pointed object can be used to punch holes in the paper about as round in size as is an ordinary pencil.

Punch holes in the paper about 2 or 3 in. apart and in rows spaced 3 or 4 in. apart. The intention is to perforate the paper so water could easily pass through. An excessive number of holes is neither needed nor desirable.

Roll the perforated paper to form a tube with the shorter dimension being the axis. Place the tube into the prepared percolation test hole. Then open the tube until it softly fits the sidewalls of the hole. The size of the bag should have been chosen so that the sidewalls of the hole are covered with considerable overlapping of the paper upon itself. The length of the paper tube may have to be cut to size, so that the tube rests on the almost flat bottom of the hole and protrudes perhaps a couple of inches, or so, above the top of the hole. Lastly, roll the top of the paper back to form a stiffened collar and to hold the overlap in place.

Fold the bottom of the bag which had been first cut out, over itself once. Then fold it sideways such that the first fold lays upon itself. Holding the folded bag bottom at the corner of the folds, cut a quarter circle at the outer edge such that a full circle is had when the bag bottom is unfolded. Choose a radius of the quarter circle about an inch or so larger than the radius of the circular paper tube in the test hole. Then place the circular bag bottom at the test hole bottom. Unfold it within to get it inside. The bag bottom with a larger diameter than the paper tube lining the hole should be pushed into place at its outer edges to fit well.

Lastly, place gravel over the bag bottom at the bottom of the hole such as to hold the paper in place when water is added later. Only an inch or so of gravel is needed. Sometimes a flat rock serves when gravel is not at hand. If one wishes, he could use a 2-in. layer of fine gravel or coarse sand instead of using the bottom of the paper bag at the hole bottom.

When a large paper bag is not on hand and the test holes are only to be perhaps 6 in. in diameter, or so, newspapers can be used instead to form a paper basket inside the hole. A double thickness is needed.

FLOAT GAUGE

Data to be presented will cause most technicians to use float gauges for measurements of water levels inside test holes. Such gauges provide needed accuracy, and they are much easier to use than tape measures, yardsticks, and the like. Float gauges are easily devised. The writer uses one as follows.

A metal rod, ½ in. in diameter and about 18 in. long, has a sharpened end. The rod is driven into the ground beside small test holes, or into the outer edge of the inside bottom of a wide test hole. Because of use of the chosen clamp, the rod is most conveniently driven at about a 30– to 45-degree angle off of vertical, and it is to one side of the hole center. Although any metal will serve for the rod, the writer has preferred aluminum alloy rods sold in chemical supply houses for support frames, e.g., Structo-Frame Rods or Flexaframe Rods.

A clamp is next connected to the rod. The clamp connects in any angle in any plane. It is: clamp holder, right angle, swivel type, Item D 1756, A. Daigger Co., 10 Harbour Blvd., Richmond, California 94801; or, contort clamp, from either VWR Scientific located in several cities, as Cat. No. 76, Item 21661-000, or from Fisher Scientific Co. located in several cities, as Cat. 77 Item 5-764.

Next a clear, rigid, plastic tube about ½ in. in diameter and 14 in. long is clamped vertically over the test hole. Graduations on the plastic tube are provided by an adhesive, plastic tape printed in inches and graduated in tenths of an inch. That tape is available as Scalafix Scales, Made in England, distributed exclusively in the U.S. and Canada by BEL-ART PRODUCTS, Pequannock, New Jersey, Item No. H-13377, and may be obtained through A. Daigger Co. or VWR Scientific.

A styrofoam ball about 2 in. in diameter serves as a float. A hollow brass tube, about 3/16 in. in diameter (from hobby or hardware shops), about 24 in. long, is pushed into the float ball. The float rod is inserted into the vertical plastic tube from below. Then the graduated tube is adjusted up or down in its clamp and fixed so that the upper end of the float rod will be within Scalafix graduations when the hole is either empty or filled with water.

ADDING WATER TO THE HOLE

Water supplies usually must be taken to the percolation test site. Convenience is important, and clean water is essential. Use clean 5-gal plastic jugs, each with a screw-capped opening and a separate small hole for releasing air locks. They can be hand carried a distance, being quite portable. A few jugs can constitute a sizeable water supply, and they can be conveyed to the site by ordinary automobiles. They are superior to commmonly used, dirty, heavy, 55-gal oil drums, which must be mounted on a truck and cannot be carried to a place inaccessible by truck.

With the paper basket in place, water can be carefully poured from the jug into the hole. Where for some reason a paper basket is not in the hole, a flexible hose should be connected to the jug and water eased into the hole through the hose. By pinching the hose, water flow will be gentle and only fast enough to fill the hole faster than water seeps away. The soils under no circumstances should be exposed to a direct flow of water such as to churn them. Break the flow on a rock or by cupping the hand over the hose end.

An expensive but most manageable hose is 5/8 in. in inside diameter with 1/8-in.-thick wall, called Amber Latex tubing, from chemical supply houses. A rubber stopper fits into the water jug's screw-capped opening, and the hose is slipped onto a piece of ½-in. copper plumbing tubing which has been fitted through a hole in the rubber stopper of the jug. A hole must be had to prevent airlocking.

If performing a USPHS test, it is often convenient to water fill the test hole to the mark at the onset of presoaking. Measure the drop in water level during the first 5 min. Then fill the hole with waters for routine presoaking. The 5-min.-presoaking water drop, as a coarse rule of thumb, tends to be about the drop that will be measured during the last ½ hr of the 4-hr test period on the following day. Not always, but often enough, that rule of thumb helps to guide the technician in his plans for time and water amounts needed for the test on the following day.

THE MINI-PERC

When preparing a percolation test program, it is helpful to know about how fast water supplies will dwindle. Water supplies must generally be brought to the site, and the amount of water needed is related to the rate at which it will be used.

Sometimes it is helpful in viewing soils in profile to be able to predict chances of percolation tests passing a locally set limitation on acceptable rates. In addition to experienced visual judgments, which often are about equal to percolation tests, the writer has found that a small bottle of water with an eye dropper can be used as follows for subjective judgments.

Break a clod of the subject soil open to expose a face of fresh soil. From a height of about 1 in., cause 1 drop of water to fall onto the horizontally held soil face. The water drop will spread, approximately wet a circle, and infiltrate. Before infiltration, the water surface will glisten. At the exact time of complete infiltration, the glistening will suddenly give way to a dullness. When the soil is rapidly permeable, the wet circle will seem scarcely more than the drop in diameter, and only a fraction of a second will pass before infiltration is complete. When a slowly permeable soil is in hand, the water drop will wet a much larger circle, and the time required for infiltration will be on the order of several seconds. If the soil is about as permeable as a rock, the water drop will spread for quite some time over the open face of the clod, and infiltration will not be a sharply defined event.

The mini-perc likely would lend itself to quantitative measurements, but the writer has neither performed the developmental studies needed nor plans to do so. He has performed the mini-perc at many sites where he also performed percolation tests. Thereby, a subjective experience has been gained which helps to predict the outcome of percolation tests to come.

A NOTE

The question is often asked, "Why such care in the percolation test, considering the soil destructive actions of a backhoe digging a disposal field? After all, the percolation test seems to be a freshwater simulation of a disposal trench."

The question is answered by the word "empiricism." The test is empirical. A short-term, fresh-water test cannot simulate the long-term, wastewater infiltration of a disposal field. And it would be technical nonsense to use a soil-destructive percolation test to guide soil-destructive construction practices. The effort should be percolation tests performed to obtain correct measurements. And the effort should be improvement of construction practices to reduce soil damage. The biggest step in the latter direction would be to dig soils for disposal fields of a septic-tank system during drier seasons of the year.

RYON'S PERCOLATION TEST PROCEDURE IS RECOMMENDED

Henry Ryon's test is simpler in procedure than the USPHS test and most other varieties of tests. Ryon's test, being more logically oriented, does not suffer strange behaviors the illogical USPHS test suffers, such behaviors to be explained in Chapter 4. Still, more exact wording is needed for Ryon's test, inasmuch as Ryon's simple wording has escaped understanding in many minds.

General

It is assumed that the technician understands careful techniques of working with soils, discussed above, and uses them in performing the test procedures to follow.

Preparing the Hole

Remove soils overlaying the stratum of concern such that a technician will have space needed to perform the test within that stratum.

After having brushed loose soil to one side, hand-dig the test hole to either 12 in. square or about 15 in. round. The bottom of the test hole should be 7 or 8 in. below the working surface of the exposed soil stratum.

Place a perforated paper basket into the hole with the basket bottom held in place with an inch or two of clean gravel or another clean weight.

Fit the hole with a float gauge such that when 6 in. of water are over the soil bottom of the hole the technician will be able to ascertain that exact water level with the gauge. Where soils are to be presoaked overnight and it would be undesirable to leave the float gauge in place, make whatever measurements or preparations are needed to fit the float gauge properly at the time of the test.

Presoaking

The purpose of presoaking is to cause soil colloids with a shrink-swell potential to swell. If such colloids are present in quantities enough to result in extensive cracking of dry soil, perform the test only when such cracks are swelled shut. When doubt exists, or when a clear understanding of the pertinent aspects of soil science is not had, presoak.

Presoaking is accomplished by water filling the hole once, or twice, and leaving the hole overnight. Where large cracks in the ground are had, extensive presoaking or a season of rains may be needed to shut the cracks.

Test Measurements

Water fill the hole to exactly 6 in. deep over the soil bottom. (Do not regard gravel or the like holding the paper basket in place as the soil bottom.) Record the time required for the water level to recede to 5 in. deep. As soon as possible, and not letting more than a minute or two pass, refill to 6 in. deep. Again record time needed for water to recede to 5 in. deep. Continue this process until successive time measurements agree within 10%. The last measurement is the percolation rate, min/in.

SOILS WITHOUT MUCH COHESION

Some soils have little cohesive strength, and a percolation test hole in them will not tolerate water fillings without collapse of sidewalls or migration of soils into the hole.

Collapsing Sidewalls

Collapse can most times be prevented by the paper basket described. It is surprising how much support paper can offer. As a matter of record, the paper basket was devised to prevent collapsing of test holes in a soil especially prone to do so. Finding a reduction in damage from water fillings was a happy benefit.

Unaware of the use of paper baskets for support of sidewalls of percolation test holes, some technicians place a length of 4-in. diameter, perforated, plastic pipe into a test hole. Then they fill the annular space between the pipe and the sidewalls of the hole with gravel. The test is performed inside of the pipe.

In some places authorities prescribe pipe and gravel as routine. Unfortunately, many people do not realize that water levels in such gravel-assisted holes recede faster than in the same-sized holes without gravel and no corrections of overly liberal results are made.

Rather than undergo the mathematical considerations to account for presence of gravel, a technician might prefer to construct a hole about 2 in. or so larger than the finished hole was to have been. Then he might fit a wire basket, or the like, of the intended hole dimensions inside of the larger hole. Small gravel could then be placed between the basket and soil wall of the hole. Although gravels and soils differ in porosity, it would be sufficiently accurate for most purposes to assume porosity of the gravel the same as that of the soil, and therefore the inside dimensions of the basket are essentially those of the originally intended test hole.

Soil Migration

Soils with large amounts of silts and few clays are prone to migrate into the bottoms of water-filled holes. Whereas collapse of a sidewall is obvious, soil migration is not. It is observed when a technician returns the next day after presoaking and he finds the test hole is not as deep as it was the day before. When silts are found to migrate, a special construction technique will be needed for both a percolation test hole and the disposal field to follow (Chapter 14).

A square hole could serve, but a round hole might be easier to manage. Say the originally intended test-hole size was to have been about 15 in. round. Dig the hole to about 21 in. round. Construct three sheet metal rings, 19 in., 17 in., and 15 in. in diameter and longer than the hole is deep. Perforate only the 15-in. ring with holes about 1/8 in. in diameter, spaced about 2 in. apart, and in rows about 2 in. apart. Place the rings concentrically within the hole. Fill the annular spaces with fine sand next to the soil, coarse sand in the middle annular space, and finally fine gravel in the annular space of the inside ring. Carefully lift out the two outer rings. Perform the test within the remaining, perforated, inside ring.

The intention of the foregoing is to provide a fine sand facing the soils to keep silt from migrating to the inside of the hole when waters are later added.

SOIL PERMEABILITY

The empiric percolation test is related to soil permeability, but it does not measure it, as such. The objectives of the percolation test are decisions on acceptability of the soil for septic-tank practices and design sizes of disposal fields. Once septic-tank effluents have infiltered a soil through a man-made device, the disposal field, the ability of the soil to conduct the effluents away, soil permeability and hydraulic gradient are generally ignored.

Unlike percolation tests which yield values related to specific procedures followed, measurements of soil permeability properly performed yield same values for same soil samples. Generally, rates are expressed in cm/sec or ft/day.

In his text [6], Professor Laak presented design criteria for disposal fields related to direct measurements of soil permeability. Laak described a falling-head permeameter and a pit-bailing test for assessment of soil permeability. Cedergren discussed several methods in detail [28].

At least in California, persons qualified either to perform or to judge the value of measurements of soil permeability are neither common nor much involved with septic-tank practices. Where regulations specify a percolation test, local authorities are acquainted with meanings of results which they must process. A pit-bailing test is not a common field practice. However sound it may be, local authorities are not prepared to interpret its results. Thus, local authorities have refused pit-bailing tests from the writer, and the use of his falling-head permeameter has provided data meaningful only to himself. Many western soils are so stone filled that valid samples for permeameter tests cannot be collected. And clients more willingly support required percolation-test programs than measurements of permeability which are not required by authorities.

Table III. Comparison of Values of Coefficients of Permeability, Percolation Rates, and Descriptions of Porous Media[a]

Percolation Rate (in./hr)	Permeability (cm/sec)	Soil Type	
	10^{2}		
	10^{1}	clean gravel	
	10^{0}		
	10^{-1}	clean sands, clean sand and gravel mixtures	
60	10^{-2}		"impervious" soils modified by effects of vegetation and weathering
14	10^{-3}	very fine sands, organic and inorganic silts, mixtures of sand silt and clay, glacial till, stratified clay deposits, etc.	
3.1	10^{-4}		
2.0	10^{-5}		
0.16	10^{-6}		
	10^{-7}		
	10^{-8}	"Impervious" soils, e.g., homogeneous clays below zone of weathering	
	10^{-9}		

[a]After Casagrande and Fadum [28], except for percolation rates calculated by a formula of correlation [29].

The writer has contented himself by correlating the USPHS percolation test to measurements of soil permeability [29]. Thus, a required percolation-test program can be performed to satisfy local demands, and data can be translated by correlation to, k, Darcy's coefficient of permeability. That value can be used to evaluate the assimilative capacity of a site (Chapter 9, this volume).

At a correlation coefficient of 0.75, paired data from 34 sites yielded a formula (corrected from earlier publication):

$$\log k = -4.76 + 1.55 \log p$$

where Darcy's coefficient of permeability, k, has units of cm/sec; and the USPHS percolation rate, p, has units of in./hr. The USPHS test must be performed in a 4-in.-diameter hole, and techniques of hole digging and water filling described herein must be followed for the formula to hold.

To provide a reader with a scale of reference, Table III is presented. The coefficients of permeability and descriptions of porous media are from A. Casagrande and R.E. Fadum, well known names in soil mechanics [28]. The corresponding percolation rates are from the formula of correlation just described.

CHAPTER 4

REALITIES OF PERCOLATION TESTS

The percolation test has a poor reputation, and little wonder. Not only are poor techniques authoritatively advised, but the ordinary technician adds his carelessness too. Even engineers trained in matters of hydraulic head have performed what they believed were proper U.S. Public Health Service (USPHS) percolation tests in holes drilled 30 ft deep in fractured granite. Brimfilled with water, those holes with 30 ft of water in place of the prescribed 8 in. of water depth were reported passing. Unusual? Not at all. Carelessness has been the rule.

One high-ranking California official viewed the writer's techniques for performing percolation tests. He was shocked to see how receptive soils really were to water. His entire frame of reference, his experience with commonly performed percolation tests, needed enormous adjustment. He muttered something to the effect, "You should have a percolation test handicap in this state."

The percolation test has many technical weaknesses [30], and no one could have looked down on the test more than has the writer [31]. Nevertheless, years of field experiences with varieties of percolation tests have caused a change of attitude. And there is the real world. Many, if not most, places in our country can barely afford the percolation test. Anything more complex, regardless of its technical merit, simply would not be practical. Further, most shortcomings of the percolation test come from ignorance of proper techniques for its performance.

There are several influences on rates at which water seeps into holes in the ground. The more salient are described herein, along with miscellaneous information which should be of interest.

DIFFERENT TECHNICIANS GET DIFFERENT RESULTS

Often local authorities know that different technicians get somewhat different results from percolation tests, but they do not have much feeling for how different. On occasion, a local authority has performed his own tests to check a technician in the private sector. Test results have been so different as to cause accusations of dishonesty to be made. The culprit might be a dishonest or careless technician, but more often it is unappreciated realities of the percolation test itself. During work on a subdivision in Riverside County, California, circumstances provided an opportunity to document these assertions [32].

In Riverside County, Ludwig's Modification of the percolation test was standard practice of local soils engineering firms. Hole diameter, depth of water-fillings, everything was rigorously set, except how the hole was to be dug and how water was to be added. The situation offered an excellent opportunity to measure differences in personal skills of technicians.

Nine locations were chosen with 144 ft being the distance between locations farthest apart. At each location, three percolation tests were performed within 20–33 in. of one another. The writer's technician, Firm No. 1, and technicians from local engineering Firms Nos. 2 and 3 each performed one of the three tests at each of the nine different locations and on different days. One test did not affect the others nearby.

As shown in Table IV, results obtained by one technician did not resemble those of other technicians. The high and low values had no tendency beyond chance to occur at the same locations. The arithmetic mean values of the three technicians from Firms Nos. 1, 2, and 3, respectively, were 23, 53, and 113 min/in. Those mean rates indicated a

Table IV. Comparisons of Technicians' Skills[a]

	Percolation Rate (min/in.)			Hydrometer Analyses (%)			
Location	Firm No. 1	Firm No. 2	Firm No. 3	Sand	Silt	Clay	Texture
1	23	118(h)	37	72(h)	14(l)	14	sandy loam
2	15	59	172	58	29	13	sandy loam
3	2(l)	15	32	41	38	21	loam
4	130(h)	91	161	46	37	17	loam
5	5	59	73	45	35	20	loam
6	2	12(l)	24(l)	52	38	10(l)	loam, sandy loam
7	3	39	229	23(l)	51(h)	26(h)	silt loam, clay loam
8	4	22	34	57	28	15	sandy loam
9	19	63	259(h)	44	36	20	loam

[a]h = high value; l = low values

very good soil, on one hand, and a very poor soil on the other hand. The lowest measured value of 2 min/in. and highest of 259 min/in. would cover the range of results one would expect of most of the world's soils.

There is a need for technicians to obtain like percolation test results from like locations, and the directions offered in Chapter 3 (this volume) would go farthest towards that end.

A COMMON ERROR IN AVERAGING USPHS TEST RESULTS

The USPHS *Manual of Septic-Tank Practice* [15] advises, "Six or more tests shall be made in separate test holes spaced uniformly over the proposed absorption field site." Despite that advice, most local authorities demand far fewer tests than needed. They often use the arithmetic mean (average) of data in deciding whether or not to accept a lot for a septic-tank system or for sizing a disposal field.

The USPHS test measures inches of water level drop during the last 30 min of a 4-hr test period. Properly reported, the units would be in./hr. But the *Manual of Septic-Tank Practice* uses the reciprocal min/in., and most local practices conform. In those places, the technician reports his calculated min/in. rather than his measured in./hr. Then unrealized errors come from averaging his data expressed as min/in.

Not everyone has close memory of his mathematics and a rule is forgotten: the average of a group of numbers is not equivalent to the reciprocal of the average of their reciprocals. For example, say the technician measured 5 in. of water drop in one hole and 15 in. in another hole during the last 30 min of the USPHS 4-hr test. Those rates would properly be 10 in./hr and 30 in./hr, and the average would be 20 in./hr. According to many local practices, however, the technician would calculate 10 in./hr as being equal to 6 min/in. and 30 in./hr as being equal to 2 min/in., and report the latter values. Then, if the reported 6 min/in. and 2 min/in. are averaged, the result would be 4 min/in. But 4 min/in. is only equivalent to 15 in./hr, which is not the proper average of 20 in./hr.

If a local authority prescribes the USPHS test, practices in his area should use in./hr for units. Otherwise, everyone should remember that it is improper to average min/in. It would be necessary to have all data in in./hr, average those data, then convert the average in./hr into min/in. for whatever use the average figure is put to.

Ryon's test, which was recommended over the USPHS test, yields min/in., and those results can be averaged validly.

CENTRAL TENDENCY OF REPLICATED PERCOLATION TESTS

In following discussions of central tendency of curves, the arithmetic mean equals the sum of values divided by the number of values. The geometric mean is calculated by summing the logs of each value, dividing the summation by the number of values, and then taking the antilog of the quotient. The median is the middle value when the values are arranged in order of magnitude. The value 4 is the median of values 1, 2, 4, 7, and 7. The mode is the value that occurs most often in a group of values large enough to have some same values. The mode of the foregoing series of numbers is 7. Usually, values must be grouped into ranges or classes for a mode to be discerned.

Studies of Replicated Tests

In 1954, investigators of the USPHS [33], drove a 4-in.-diameter cylinder 3 in. deep into soil in the bottom of a freshly dug disposal trench. A "confined" percolation test was performed within the cylinder. This was replicated 24 times at evenly spaced locations along trenches of 2 adjacent, new systems within an area of about 0.2 acre. Results have been entered in Table V as Study No. 6. (USPHS studies comparing the foregoing confined percolation test to the unconfined ordinary USPHS percolation test, were, by the writer, found to show the confined test rate to be equal to 3/4 of the unconfined test rate in in./hr.)

Table V. Central Tendencies of Replicated Percolation Tests

			Percolation Rates					
					Mean			
Study No.	Number of Replications	Test Procedure	Units	Range	Geometric	Arithmetic	Median	Mode
1	36	USPHS	in./hr	2.6–78	11.7	15.39	11	11
2	62	USPHS	in./hr	0.6–150	6.0	12.13	5.6	3.7
3	6	USPHS	in./hr	3.3–8.6	5.2	5.48	4.9	4.55
4	20	USPHS	in./hr	0.0–11.5	1.68	3.47	3.05	0.1
5	31	USPHS	in./hr	0.13–24	1.27	3.92	2.0	0.2
6	24	USPHS	in./hr	0.1–4.2	1.2	1.91	1.85	0.1
7	20	USPHS	in./hr	0.08–6.5	0.68	1.58	0.45	0.4
8	24	USPHS	in./hr	–0.8[a]–4.0	0.51	0.85	0.32	0.1
9	42	Ryon	min/in.	0.32–125	7.5	22.73	5.0	<1
			(in./hr)	(0.48–188)	(8.0)	(2.64)	(12)	(>60)
10	6	Ryon	min/in.	4.29–120	24.2	47.16	29.7	5.6
			(in./hr)	(0.5–14)	(2.48)	(1.27)	(2.02)	(10.7)

[a]Negative readings are occasionally had, e.g., from soil sloughing into the hole.

During percolation test programs on subdivisions, occasionally many tests were performed in a same soil type, identified as such either by a highly qualified soils scientist or by mechanical analyses of soil texture. With the exception of the Study No. 6, all studies in Table V were performed by the writer and his technicians. Techniques were careful and consistent.

In Table V, Study No. 1 was performed in a quite permeable sandy loam in Kern County, California [34]. Study No. 5 was performed at the same location, but tests were in holes (often constructed with mallet and chisel) in weathered granite.

Study Nos. 2, 3, 9, and 10 were performed in quite permeable topsoils in San Mateo County, California [35,36]. Study Nos. 7 and 8 were in subsoils about 5 ft deep, and Study No. 4 was in subsoils about 10 ft deep in San Mateo County.

The USPHS Test

To assist with interpretations, data from the studies of the USPHS test were normalized by dividing each study group by its high value, multiplying by 100, and then data were combined into groupings displayed by the histogram in Figures 6 and 7. In Figure 6, the 90.1–100 bar was omitted because normalization exaggerated that range grouping.

It can be seen that replicated test data strongly tended to be slower rates. Even when the 0–10 range was expanded in Figure 7, the same strong tendency was displayed for the curve to be skewed towards slower rates.

Ryon's Test

Data from 42 replicated tests of Study No. 9 were grouped and presented in Figure 8. As before with USPHS tests, data tended to cluster at an extreme. But as opposed to the USPHS test, Ryon's replicated tests were strongly skewed towards faster rates. The expansion of values in the 0–9.9 range in Figure 9 also showed the same trend towards faster rates.

Central Tendency

The arithmetic mean, more commonly referred to as the average, does not constitute a good description of central tendency of either the USPHS test or Ryon's test. A few extreme values in a skewed curve always pull the arithmetic mean far to one side of a central grouping. Note Figure 8: the arithmetic mean (22.73 min/in.) misses most of the data points.

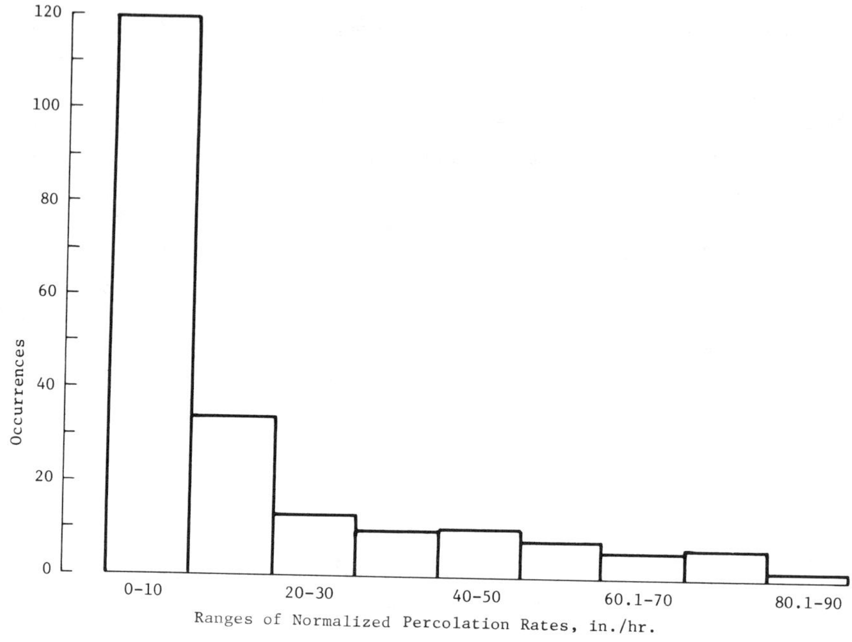

Figure 6. USPHS percolation test procedure, combined normalized rates from Study Nos. 1 through 8, Table V, each study of replicated tests having been in a same soil type.

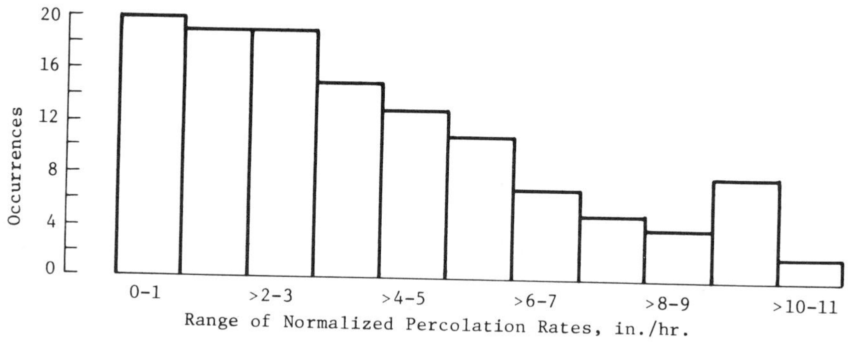

Figure 7. Expansion of the 0–11 range of values combined as a single histogram bar and a portion of another in Figure 6.

As shown in Table V and Figure 9, the mode is descriptive of only the fastest Ryon rates.

The geometric mean seems to describe best the central tendency of replicated percolation tests.

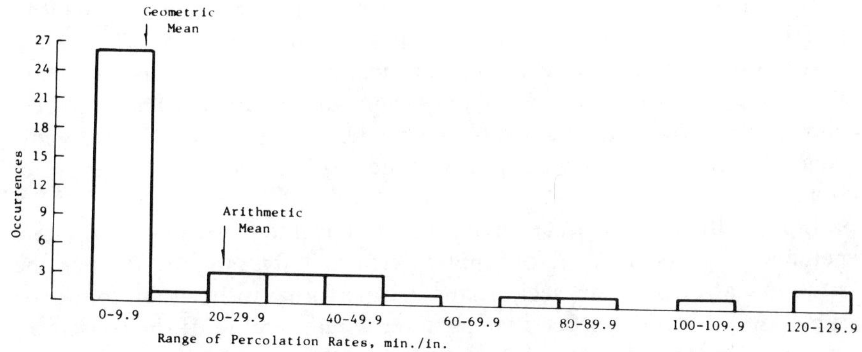

Figure 8. Ryon's percolation test procedure, 42 tests of Study No. 9, Table V.

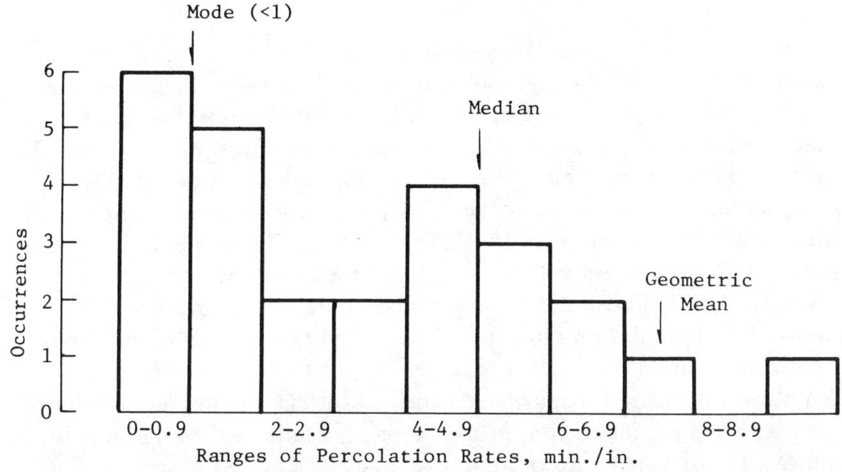

Figure 9. Expansion of the 0–9.9 range of values combined as a single histogram bar in Figure 8.

Suggestions

The field data do not support an early statement of investigators of the USPHS [33], ". . . use the average rate (arithmetic mean) as the best estimate of soil absorption characteristics." The arithmetic mean is too far from most replicated data to describe a central tendency well.

In a subdivision situation it would seem best practice first to identify soil types, soils being the particles passing a 2-mm sieve [37]. Then percolation tests could be performed in multiplicity until an adequate histogram of results from each soil type could be drawn. Then the best choice of central tendency could be made. The geometric mean seems generally best, but practical circumstances might dictate another selection.

In an individual lot situation, it is not practical to perform a great many percolation tests. Rather, one must perform only as many tests as the situation affords. Some technicians can perform five tests in about the same away-from-the-office time as fewer would require. If the technician can solve for the geometric mean, he is advised to do so. If not and if five test results are had, he is advised to ignore the two slowest and one fastest of Ryon's tests, or to ignore the two fastest and one slowest of the USPHS test, and then to average the two remaining tests.

Fallacy of Percolation-Rate Maps

In some localities, authorities demand from a percolation test program a map of the subdivision giving soils delineated as to percolation rate ranges. Those maps are not constructed by studies as suggested above; rather, percolation tests are scattered over the subject terrain. Then the engineer draws lines around similar results, and he so designates areas where all soils are believed to be within a percolation rate range. Such maps resemble soils maps provided by the U.S. Department of Agriculture Soil Conservation Service (USDA-SCS) in appearance.

See Table IV. Those studies were made on a single lot of a proposed subdivision. Percolation rates from 2 to 259 min/in. were measured. Soils contained from 23% to 72% sands, from 14% to 51% silts, from 10% to 26% clays, and varied from sandy loams to silt-clay loams. A percolation rate map of that subdivision, properly made, would have had about every square foot of soil to account for.

For another example among many that could be offered, see Table V. Study No. 6 was performed within an area of 0.2 acre. In terms perhaps more familiar to most readers, rates of 0.1 to 4.2 in./hr would amount to 600 min/in. to about 14 min/in.

Subdivision engineers provide percolation rate maps giving a soil with this rate range here and a soil with that rate range there, because it is demanded of them. Those engineers honestly try to conform, and they likely believe their maps accurate. But if they performed special studies such as were performed by the writer [32,36], their percolation rate maps would likely be discredited.

REPRODUCIBILITY OF TESTS

Most percolation test programs, carefully performed, yield a wide array of results. One might believe that soils varied accordingly, or one might suspect the test of being highly variable in itself.

During October, USPHS tests were performed at 20 locations on a proposed subdivision, care being taken in techniques of digging, water filling, etc. Results varied from 1.0 to 9.8 in./hr (6.1 to 60 min/in.). The arithmetic and geometric means, respectively, were 4.85 in./hr (12.4 min/in.) and 3.95 in./hr (15.2 min/in.).

During December of that same year, tests were again performed at those 20 locations, in holes within a few feet of prior test holes. Results varied from 0.6 to 35 in./hr (1.7 to 100 min/in.), and there was greater difference between test results at a same location than between locations. At one location the two tests were only 0.6 in./hr different, but at another the difference was 26.8 in/hr. The paired tests only had a correlation coefficient of 0.38, statistically of low significance.

Finding very different results between two tests at a location would be in line with experience. Soils vary greatly in permeability, even soils only a few feet apart.

Nevertheless, the arithmetic mean rates in October and December were 4.85 in./hr (12.4 min/in.) and 5.98 in./hr (10.0 min/in.). Those arithmetic means were about 21% apart. The geometric means, better measures of central tendency of those respective months' tests, were 3.95 in./hr (15.2 min/in.) and 4.22 in./hr (14.2 min/in.). Those geometric means were less than 7% apart. Within error expected of such a field experiment, the agreements were good. The percolation test performed in accordance with instructions given herein concerning care seems very reproducible.

DECELERATION OF WATER LEVEL DECLINING IN A TEST HOLE

When water is added to a hole in soil, it infiltrates at a decelerating rate. Inasmuch as the USPHS percolation test reports a velocity measurment, just where and when measurements are made of a deceleration will affect the outcome.

A Typical Infiltration Curve and Its Meaning to the USPHS Test

Eight holes were kept filled with water until saturation was complete. Then water-fillings were stopped. As the declining water levels passed successive inches downwards in each hole, time was recorded. The holes behaved essentially alike, so data were combined to draw Figure 10.

The *Manual of Septic-Tank Practice* [15] directs, "In sandy soils (or other soils in which the first 6 inches of water seep away in less than 30 minutes, after the overnight swelling period), the time interval between measurements shall be taken as 10 minutes and the test run for one hour." Otherwise, the time interval is directed to be 30 min for a 4-hr period.

Note Figure 10. During the first 10 min, water level dropped 2.7 in., or at a "velocity" of about 16.2 in./hr (3.7 min/in.). But during 30 min., the water level only dropped 5.2 in., or at a "velocity" of about 10.4 in./hr (5.8 min/in.). One would like to believe the USPHS test would yield same results, whether or not a patient technician chose to refill a hole until 30-min readings could possibly be made. But same results from 10-min and 30-min intervals are not obtainable. It is illogical to hold time a constant while permitting changing physical circumstances to vary, and then to calculate a velocity to describe deceleration. Thus the USPHS found it necessary to change "the constant time" of 30 min to 10 min to fit circumstances "beyond control."

Note in Figure 10 that the USPHS 8th inch infiltered at 2.5 min/in. (24 in./hr), but Ryon's 6th inch infiltered at only 5 min/in. (12 in./hr). Only 2 in. of water filling over 6 in., or one-third more, would appear to double infiltration rates.

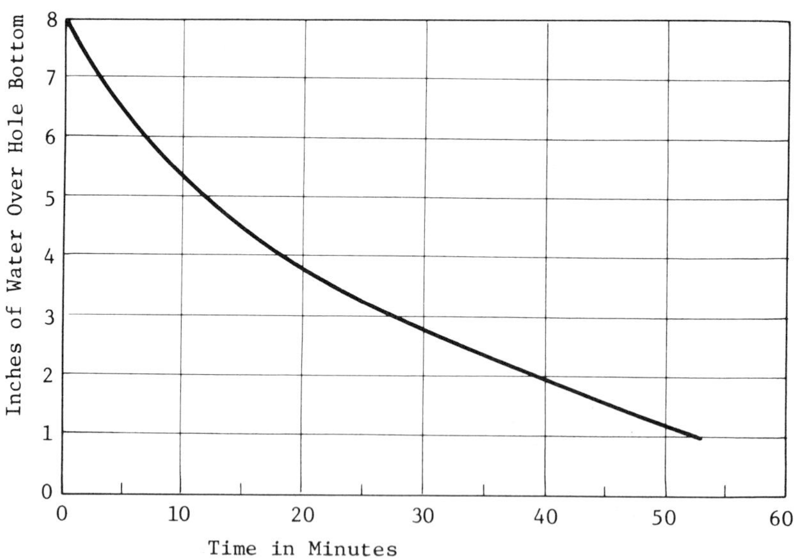

Figure 10. Typcial curve of receding water level into a percolation test hole.

EFFECT OF TIME BETWEEN REFILLINGS

Depending on how fast a water level declines, soil left above the declining water level drains and rests until the test hole is refilled.

In a rather permeable sandy loam, a customary 4-hr test was performed. Refilling and measurement intervals had been 30 min each. The percolation rate of the 30-min interval test was R^{30} = 12.6 in./hr. Immediately, the intervals between refillings and measurements were shortened to 10 min each, and the stabilized rate became R^{10} = 18.3 in./hr. Similarly, successive stabilized rates of shorter intervals were: R^{5} = 21.6 in./hr; R^{2} = 24.0 in./hr; and R^{1} = 24.0 in./hr.

It would appear that the successively shorter time intervals resulted in greater permeabilities. Such was not really the case. Because shorter time intervals placed measurements towards the steeper part of the curve of deceleration (see Figure 10), faster velocities were being calculated. Actually, the rapid refillings shortened the resting of upper soils of sidewalls, and they were becoming less receptive to infiltration than before. Immediately after the last measurement of 1-min intervals, R^{1} = 24.0 in./hr, the schedule was returned to a 30-min interval. The first interval yielded R^{30} = 6.2 in./hr. That was a sizeable reduction in "stabilized" rate, its being about a fourth of what it had been immediately before. But the rest of upper soil surfaces during that first 30-min interval following 1-min intervals had its beneficial affect. The second interval was R^{30} = 6.3 in./hr. Rates were returning to the fast rate, R^{30} = 12.6 in./hr, which was the stabilized rate of the 30-min interval at the onset of the study.

The data demonstrated what was experience. When the sidewall of a hole is inundated with water, infiltration rates decrease for a while. When allowed to drain and rest (unless damaged by careless water addition techniques), that sidewall recovers some of its initial infiltrative capacity. And, as said, because of the failure of the USPHS test to recognize the deceleration of the seepage curve (which Ryon did), a false appearance of faster infiltration will be had with shorter time intervals for refilling and measurements.

As to refilling the test hole each 30 min of a 4-hr period, the *Manual of Septic-Tank Practice* [15] states, ". . . refilling 6 inches over the gravel as necessary," (8 in. over the actual hole bottom). Some local authorities either misunderstand that ambiguous directive or believe it is conservative to interpret it to mean, "refilling only if the hole threatens to run dry." That adds even more caprice to an already capricious test. As seen in Figure 10, deceleration results in far slower rates towards the hole bottom than higher up in it. It is not conservative to measure excessively slow rates. It is error. Incidentally, a well trained technician knows that refilling a test hole that "threatens to run dry" just before the last 30-min period results in faster rates than if the hole had been properly refilled each 30 min of the 4-hr period.

EFFECT OF WIDTH OF A TEST HOLE

In 1950 [33], investigators of the USPHS reported studies of effects of hole diameters, from 4 to 12 in., on percolation rates. Those studies were not adequately replicated. It was noted, ". . . the difference between rates of duplicate holes is as great as, or greater than, the difference between different-sized holes." They concluded by advising, "Use a 4-inch auger hole in which to make the test instead of the 12-inch-square dug hole."

In 1954, investigators of the USPHS [38] reported more studies with percolation test holes of different diameters. Those later studies still suffered inadequate replication. It was noted, "Actually, it appears that the differences in rates between replicate tests within the same soils are greater than the differences in average rates between different sized holes in the same soils." Without continuing to devise a definitive experiment, the USPHS produced the *Manual of Septic-Tank Practice* [15] which directs, "Dig or bore a hole, with horizontal dimensions of from 4 to 12 inches . . . "

Realizing that investigators of the USPHS were unaware of the need for extreme care in digging test holes and water-filling them, and realizing that hole size theoretically must make a large difference in percolation rates, Winneberger and Klock [32] made definitive studies in 1973. Among other test conditions, replicated narrow and wide test holes were dug with diameters of exactly 3.3 and 13 in. respectively. A gravel-filled annulus was around pipes with inside diameters of 2.1 and 10.5 in., respectively. Mean rates were 9.6 and 33 in./hr, respectively. Those studies were statistically highly significant.

Thereafter, studies of hole sizes and percolation rates were made when convenient during routine percolation test programs of subdivisions. Always, careful techniques had the same results: wider holes yielded slower percolation rates than narrower holes did.

Perhaps as a best example of many experiences, at 20 locations on one subdivision [36], a hole with a diameter of 12 in. was dug, and another of 5 in. was dug nearby, but not so close as to interfere with the first hole. Carefully performed USPHS tests in those 12- and 5-in.-diameter holes, respectively, yielded arithmetic mean rates of 0.883 in./hr (68 min/in.) and 1.58 in./hr (38 min/in.). Respective geometric means were 0.53 in./hr (113 min/in.) and 0.676 in./hr (89 min/in.).

EFFECT OF DEPTH OF WATERFILLING A TEST HOLE

When studying the percolation test [33], USPHS investigators reported, "... inherent soil differences between adjacent test holes are greater than those due to small differences in head." As for small differences in hydraulic head, it was noted, "... heads in the test holes over the last 30-minute periods of the tests varied from 2 to 6 inches."

During definitive studies in Arizona in 1973 [32], the effects of different depths of water-filling holes were shown to be significant. Mean percolation rates in holes first water-filled to 8 in. deep, then retested immediately with water-fillings 14 in. deep, were 116 in./hr, then 204 in./hr. In companion holes water-filled to 50 in. deep, the mean percolation rate was 1860 in./hr.

During percolation test studies [36], changing the depth of water-fillings from 6 to only 8 in. in one hole (Ryon's procedure except for changed depth of water), resulted in about halving stabilized percolation rates from 1.1 to 0.6 min/in. In another hole, changing water-fillings from 8 to 6 in. resulted in changing stabilized rates from 2.1 min/in. to 6.2 min/in. There, a 2-in. change in water depth, only 30%, changed stabilized percolation rates almost three-fold.

Experience with many observations of field tests lead to this assertion. The difference between water-filling a test hole only 2 in. more, which is the difference between water-fillings of Ryon's and the USPHS test, can change the percolation rate from about half again to treble. Such differences seem to be greater in the more permeable soils.

Measurements clearly demonstrate that force of seepage in a percolation test hole is not equal to the simple depth of water-filling over the bottom. It is more complex than that.

SOIL PUDDLING

When a raindrop falls on a soil surface, aggregated sands, silts, and clays are separated. There is a tendency for the smallest particles to be carried with waters infiltrating soil pores. Those fine particles eventually reduce permeability of the soil by tending to fill pores. They may form a compact layer just under the soil surface (Figure 11).

Soil puddling (and compaction) is often exhibited by persistent puddles of muddy water in roadways after rainwaters have long infiltrated unaffected soil surfaces. In most percolation test holes, soil puddling is exhibited by muddy water in the hole. Puddling results in very much slower percolation rates than would have otherwise been had.

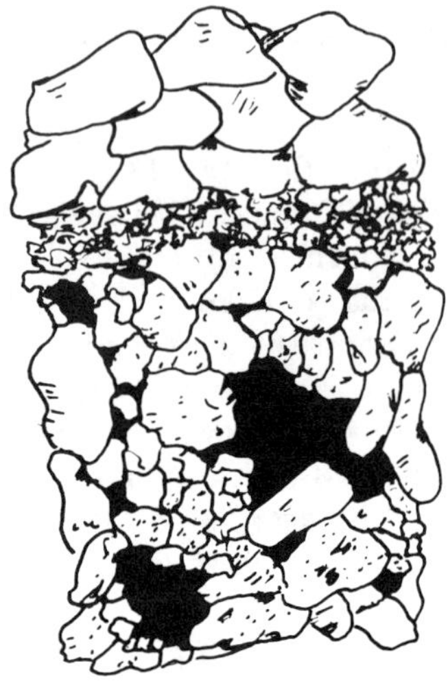

Figure 11. Soil particles aggregate and leave spaces through which waters percolate. This diagram shows the *Duley layer*, formed by rains beating on the unprotected soil surface. Fine particles migrate down and form a less pervious layer just under the larger particles stranded at the soil surface [39,40].

Soils vary in their liability to soil puddling. Sandy soils without many fine particles suffer little, but poorly cohesive clay loams readily suffer puddling.

At a location in Alameda County, California, holes were dug into Millsholm silt loam [41,42]. When water was introduced gently by a siphon hose, 4 in. of water seeped away in only 4 min. But when water was added with the hose from the hole top and allowed to splash and strike the bottom, still with less violence than adding from a bucket would have been, 4 in. of water seeped away in 80 min.

EFFECT OF A POWER AUGER

It is commonplace for technicians who perform many percolation tests to speed up the task of digging test holes by using a powered flight auger mounted on a small vehicle.

At a location where a percolation test program had been a manifest failure, measurements of soil spoil next to test holes disclosed that only 85% of the soil left the holes. The remainder of the soil was compacted into the walls of the holes. On another such occasion, there was no spoil at all. All of the soils remained compacted within.

Even if all soils were removed from the hole dug by a power auger, soils with fine particles are still seriously damaged by the violent actions of the auger. At one locality, percolation rates in three power-augered holes were >60, 80, and 240 min/in., respectively. The sides of those same holes were later enlarged with the rigid putty knife method described herein. Subsequent percolation tests in those repaired holes, respectively, were 2.4, 0.8, and 10 min/in.

The writer could continue with examples of soil-damaged percolation test holes, even holes damaged by post hole diggers and other such equipment advised by the *Manual of Septic-Tank Practice.* But perhaps the point would be strongest made by pointing out that most of the writer's income working as a consultant for many years came from clients having discharged their engineers for their failing percolation test programs. Soil-damaged test holes were not the exception. They were the rule.

A Fate of a Special System for "Impermeable Soils"

Occasionally, there arises a touted proprietary system for on-site disposal of wastewaters to what are believed to be practically impermeable soils. One of those systems enjoyed success for a while in the Southeast, but the time came when it was installed in areas where it failed.

Commonly, percolation tests are performed without techniques needed for best results. Consequently, there are vast areas in our country believed unacceptable for routine septic-tank practices because of "inadequate perc." In reality, many have an excellent capability for percolating water: witness tree growth in soils reported "no perc." Most land plants cannot grow without water and soil aeration.

The reason for the manifest failure of the particular proprietary device mentioned was its initially being installed in permeable soils which were believed to be practically impermeable. Later, soils with genuinely low percolation capacities were fitted with the system, and it failed.

COMPARISON OF RYON'S AND THE USPHS PERCOLATION TEST

Investigators of the USPHS conducted field studies of Henry Ryon's percolation test [33,38]. They believed Ryon's directions left too much subjective judgment to the technician. Thus they developed what they believed was clearer wording. And in accordance with Lloyd's law [43], "Attempts to simplify add to confusion," the USPHS *Manual of Septic-*

Tank Practice [15] presented a percolation test fundamentally different from Ryon's. Still, Ryon's empirical loading rates of septic-tank effluents onto disposal fields was used in the *Manual* with the new USPHS test. Writers of the *Manual* were unaware of the departure from Ryon's practices and its implications.

San Mateo County in the San Francisco Bay Area follows USPHS percolation test practices, except hole size has been standardized to 12 in. in diameter. Regional authorities recommend Ryon's procedures [24]. A subdivision there afforded an opportunity to perform both tests in comparison [36]. At each of 50 locations, the USPHS test was performed with water fillings 8 in. deep over actual soil bottoms of holes with diameters carefully dug to 12 in. Then from 1 to 11 days later, those holes were enlarged to 14 in. in diameter. Then Ryon's tests with their 6-in. depth of water-fillings were performed until stabilized rates were obtained.

The data, presented in Figures 12 and 13, did not delineate a straight-line relationship. Rather, a curve was visually fitted. Perusal of Figures 12 and 13 leads to the following conclusions:

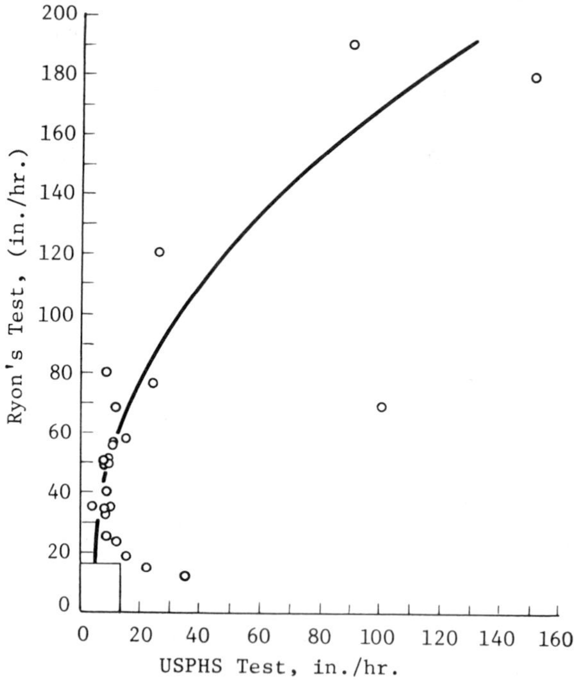

Figure 12. Ryon's and the USPHS percolation tests, performed in pairs at different locations.

1. Above about 5 in./hr (12 min/in.), the USPHS test yielded slower rates than did Ryon's test, and below about 5 in./hr, the USPHS test yielded faster rates than did Ryon's test.
2. Ryon's and the USPHS percolation tests are so basically different that results of one test cannot validly be translated into equivalent results of the other.

A SPECIAL QUIRK OF THE USPHS TEST

The *Manual of Septic-Tank Practice* [15] directs:

> 5.A. If water remains in the test hole after the overnight swelling period, adjust the depth to approximately 6 inches over the gravel. From a fixed reference point, measure the drop in water level over a 30 minute period. This drop is used to calculate the percolation rate.

The authors of the *Manual* saw no reason to delay final measurements. They rationalized that presoaking waters persisting in the hole would have saturated soils at least as much as would have been achieved during

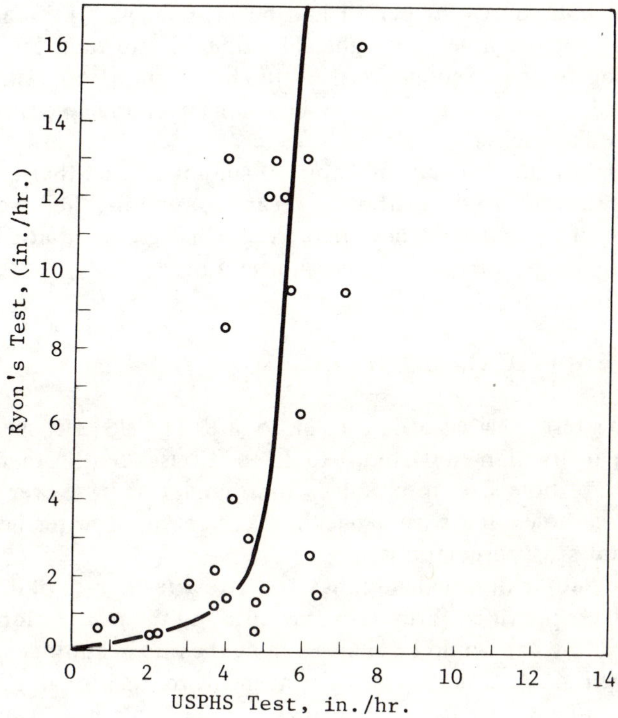

Figure 13. Expansion of small rectangle in Figure 12.

the 4-hr test itself. That would be true if the persistent waters were as deep or deeper than the prescribed 8 inches for the test. However, the technician would scarcely perform the test in such a hole where the outcome was obvious—a manifest failure to pass local rules of acceptance. So the rule becomes applied to holes with sidewalls partially or mostly drained overnight.

USPHS percolation tests were replicated ten times on a client's Lot A. Presoaking waters left overnight were found remaining in seven holes, and another three holes were found empty. All holes were tested by refilling to mark each half hour of the prescribed 4-hr period, and accurate float gauges were read each half hour.

Perusal of Table VI discloses the fallacy of the directive of the *Manual* pertaining to persistence of presoaking waters. Especially note Hole 4 which had a "final" percolation rate of 19 min/in. according to the *Manual.* But ignoring the *Manual* and performing the test for the full 4 hr resulted in a last 30-min rate of 600 min/in.

Studies of Lot A were not unusual. Turning casually through records of tests, a test elsewhere which yielded 0.5 in./hr (120 min/in.) during the last half hour of a 4-hr period had been found with presoaking waters persisting in the hole overnight. The first 30-min rate, the "final rate" according to the *Manual*, was 2.4 in./hr (25 min/in.). The client had remarked, "Gee, you mean if the hole can't even manage the presoaking water, we're lucky?"

Perusal of the statistics of Table VI supports belief that the geometric mean is superior to the arithmetic mean (commonly, the "average"), as a measure of central tendency. It offered a more conservative value with which to judge the soils of the subject Lot A.

MATHEMATICAL MODELING

Ryon's test specifies a hole 12 in. square. The USPHS specifies a hole varying in its diameter from 4 to 12 in. Consequently, there is quite a selection of hole sizes in practice. Some engineers have even water-filled very large holes dug with a backhoe and reported water level drops in those holes as percolation rates.

Some authorities, realizing that hole size affects rates of drop of water levels, have provided formulas for translating the percolation rate in one hole size to what would be believed to be the rate in another standardized hole size [24]. It would be most convenient to dig a small hole and later convert rates to what a standardized, larger hole would have yielded. Anyone carrying water to remote locations on hillsides would well appreciate the convenience of using a smaller hole.

Table VI. Comparison of Initial and Final Percolation Rates on Lot A

Hole No.	Percolation Rate: First 30 Minutes in./hr	min/in.	Last 30 Minutes in./hr	min/in.	Overnight Presoaking Waters Persisting
1	0.40[a]	(150)	0.60	(100)	yes
2	0.60[a]	(100)	0.40	(150)	yes
3	2.0	(30)	1.2[a]	(50)	no
4	3.2[a]	(19)	0.10	(600)	yes
5	1.0	(60)	0.80[a]	(75)	no
6	0.60[a]	(100)	0.40	(150)	yes
7	1.4[a]	(43)	0.60	(100)	yes
8	0.40[a]	(150)	0.20	(300)	yes
9	0.30[a]	(200)	0.20	(300)	yes
10	1.2	(50)	0.80[a]	(75)	no
Arithmetic Mean	11.1	(54)	0.53	(113)	
Geometric Mean	0.85	(71)	0.42	(142)	

[a]Official USPHS percolation rate according to *Manual of Septic-Tank Practice* [15] because, ". . . water remains in the test hole after the overnight swelling period. . . . "

Occasionally tests are performed in holes of different sizes, side by side. Calculations assuming that the rate at which water seeps into one hole size is proportional to wetted soil surfaces sometimes fit the data for the other hole size. Other times, however, they do not. As an example of not fitting, 18 USPHS tests in holes with 5- and 12-in. diameters had geometric mean rates of 0.53 and 0.65 in./hr, respectively. The slow rates only very slightly depleted the 8-in. columns of water-fillings, so aside from hole diameters, conditions were about equal in the 18 paired holes. Calculations of the volumes of water seeping away into the soil surfaces of the 5- and 12-in.-diameter holes, respectively, yielded 0.088 in.^{3}/in.^{2}/hr and 0.14 in.^{3}/in.^{2}/hr. Calculations using arithmetic means yielded only little better results, 0.23 in.^{3}/in.^{2}/hr and 0.29 in.^{3}/in.^{2}/hr (~23% apart) in the 5- and 12-in.-diameter holes, respectively.

Two matters, at least, frustrate devising a suitable model for translating percolation rates in one hole size into rates as though in another hole size—quantitatively unknown forces moving seeping waters, and unpredictable geometries of the pathways of percolation.

Seepage Forces

Capillary forces cause the blotting effect Ryon referred to when he advised presoaking the hole. Despite presoakings, many soils retain strong capillary forces around the wetted perimeter of a test hole.

Seepage forces in a test hole are unknown. The force is not the simple height of the water column in the hole. Waters themselves in soils, through tensile strength of water and various other physical phenomena, also contribute to seepage forces. For example, in one test hole, stabilized measurements of first the 6th and then the 8th inch of water resulted in rates of 54 and 100 in./hr, respectively. So far as water columns as such were concerned, the hydraulic head was increased about 1.33 times. But the rate change, which in theory might have increased proportionately from 54 to 72 in./hr was actually 100 in./hr. And in another hole, stabilized measurements of first the 8th and then the 6th inch of water resulted in rates of 28 and 10 in./hr, respectively. The 75% decrease should have changed from 28 to 21 in./hr, if depths of water columns were equivalent to hydraulic heads, but much less change was had. In those situations, whatever forces caused water to seep into holes were beyond simple considerations of the depth of water-fillings, and by unpredictable amounts which could change in different ways depending on whether or not stabilized water drops were increased or decreased.

Unpredictable Geometry

Mother Nature is not neat. Most soils are not *isotropic*, that is, they do not have the same physical properties regardless of directions of measurements. Many soils have plate-like structural units, which result in greater permeabilities sideways than downwards. Some soils, notably primitive Western soils forming in place over bedrocks, become less permeable rapidly with depth. It is not unusual for percolation test holes to have very permeable tops of sidewalls, while bottoms are into weathered rocks. Some holes are constructed with a mallet and chisel in permeable, but hard, weathered bedrocks.

On occasion, freshly used percolation test holes were cut into cross section. A portion of the variety of wetted surfaces is presented in Figure 14. As shown, a water-filled hole empties without as deep a penetration as one might believe. The patterns displayed testify to the unpredictable geometries of infiltration, which would affect percolation rates in a manner such as to frustrate efforts to treat them alike mathematically.

POSSIBLY A PERCOLATION TEST CONSTANT

Field Observations

During studies of a subdivision in 1965 [42], a small deposit of Los Osos seeped variant was found included in Millsholm silt loam. The deposit had the appearance of being almost isotropic to considerable

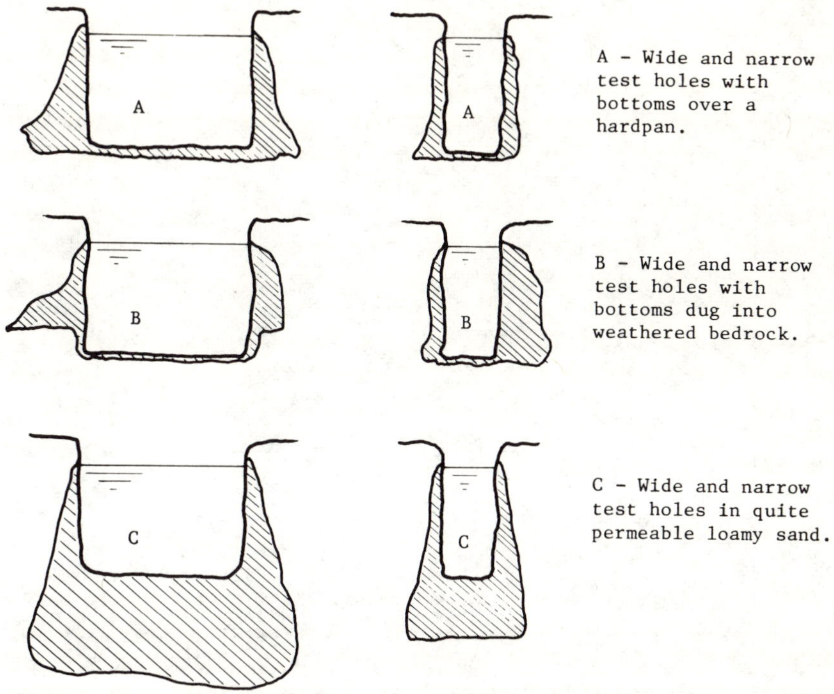

Figure 14. Some of the great variety of cross-sectional patterns of soil wetting by percolation tests. (Original level of water filling shown.)

depths. Three holes were dug, each with a diameter of 4 in. They were 1.5 ft apart, and 1, 2, and 4 ft deep. When water-filled, the deepest hole began infiltrating water faster than the shallowest did, but all waters seeped away in about the same length of time. Being a silty soil, soil migration had altered depths to 12, 18, and 23.5 in., respectively. The study was interesting, but questionable because of change in hole depths from silt migration.

Laboratory Observations

Later [41], a large box with an underdrain was filled with Oakley sand, a very permeable and structureless soil. Holes were dug 3 in. in diameter and 12 in. deep. Septic-tank effluents were dosed into those holes.

Early in studies when effluents were infiltrating the holes the same as fresh waters would, dose volumes were varied to see if consequently deeper fillings seeped away in about equal times. They did. An example of data is presented in Figure 15, where it can be seen that the time needed

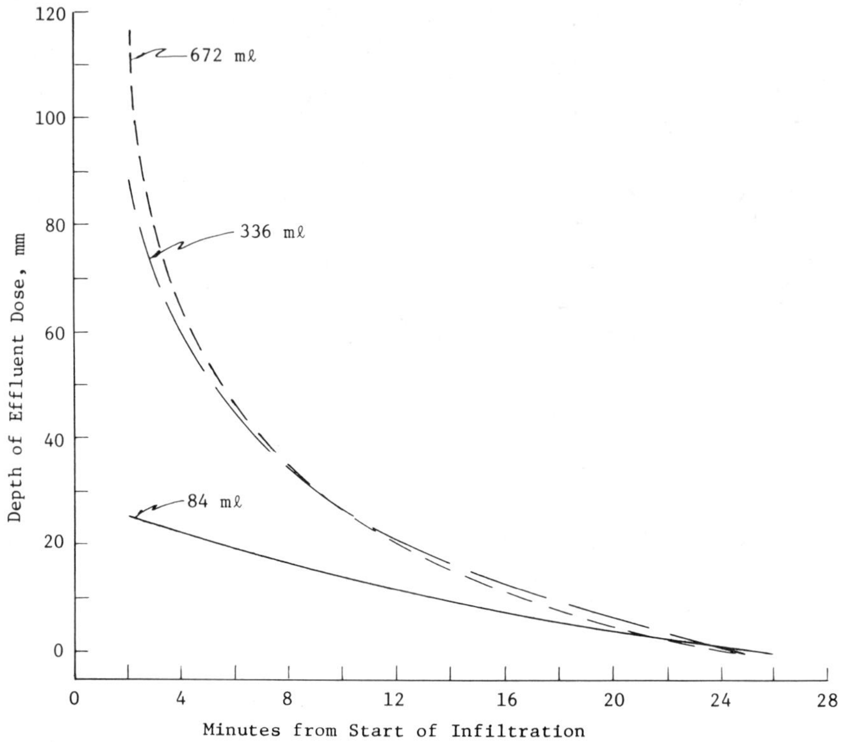

Figure 15. Infiltration of three different dose volumes of effluent into a single hole in Oakley sand.

for the doses to seep away was about the same regardless of dose volume or depth of a dose in a hole, in spite of an eight-fold difference in volume input. Hence the depth to which the hole was initially filled had considerable influence on seepage rates at any particular soil horizon. Note that when the hole was water-filled 26 mm deep (dose volume 84 ml), 24 min were required for complete infiltration. But when the hole was first filled 117 mm deep (dose volume 672 ml), the 26- to 0-mm increment of soil in that same hole required only 14.5 min for infiltration, once the surface level of the dose reached it.

The times required for complete seepage of the three dose volumes shown in Figure 15, least to most volumes respectively, were 26, 25, and 24.5 min. The tendency for a deeper filling from a larger dose volume to seep away slightly faster was tested.

Dose volumes of 143 and 286 ml were discharged into seven replicate holes each. The mean times for those respective volumes to seep away were 35 and 33 min. Statistically high significance was had. Of the seven replicates, three had same infiltration times for the different doses, four had lesser time for the greater dose volume, and none had more time for the greater dose volume to infilter.

It has been repeatedly pointed out that the forces causing water to seep into a hole are far more complex than simple measurement of the height of a water column in a hole. The foregoing measurements of seepage illustrate the point well. Nevertheless, the studies were not repeated enough times to satisfy the writer as to universal application of details. Unfortunately, those studies were not pursued further.

It is entirely possible that the length of time required for complete infiltration of a given water volume in a given hole size could constitute a reliable soil test, which might be calibrated to design size of a disposal field.

A Practical Note

When a water refill in a percolation test hole accidentally is 0.2 or 0.3 in. above the mark on a float gauge, experience has been that little or no more time is required for the extra water to seep away than had the refilling been exactly to mark. The technician must note, but not credit, the extra depth in calculations. Otherwise, he would measure a greater drop in water level per unit of time than he should.

SOURCES OF UNNEEDED, LARGE ERRORS IN FIELD MEASUREMENTS

Float gauges are the best devices for determining changes in water levels in holes. They obviate large errors which are commonplace. Most unfortunately, the *Manual of Septic-Tank Practice* and most local authorities do not specify means of making measurements of water levels in percolation test holes. Sometimes authorities wonder when field measurements are reported in nearest ¼ in. (sometimes ⅛ in.), but they soon become accustomed to the display of insensitive measurements. Other poor field techniques are probably regarded as greater errors. Indeed, they would be, but that is no reason to tolerate avoidable errors.

Reference Points

Technicians have a variety of ways to mark a place in a hole. Some use a nail and others use a twig pushed into the hole side. Those marks at odd angles scarcely provide points to which the technician can return and use accurately.

Some technicians lay a reference board across the top of the hole. Others even carry the board around with them on rounds of tests. Even when the technician leaves the board in place, it scarcely provides an accurate reference point when he returns to it, steps on it, and slides his tape measure into the hole from a standing position to read it.

Bright sunlight makes the usual hole, 2–3 ft deep or more, look very dark inside. Seeing the water surface may not be possible. Often rippling water is the signal that the end of the tape measure has arrived. It could as easily be soil sloughing from the sidewalls. At best, the end of the tape measure could be a sizeable fraction of an inch up or down. At worst, it is 2–3 in. one way or the other when water ripples for some reason.

Once a technician believes the end of the tape measure is at the water surface, he then turns his head to read the tape measure at his reference point, wherever that is at the time. Holding a tape still while moving the body is not as easy as one might think.

Water adheres to the tab at the end of a measuring tape (or ruler), if lowered for a second try. That drop of water adds at least $^1/_{16}$ in. to the effective length of the measuring tape.

Steel tapes rarely hang straight. A 31-in. length of a tape bowed out only 1 in. offers $+^1/_{16}$ in. error. If the tape is held only 3 in. off of absolutely vertical, about $+^1/_8$ in. error is had.

Anyone who has examined many field notes of technicians will recall measurements being reported to the nearest ¼ in. Thus, in many, if not most field situations, the error in reading graduations to the nearest ¼ in. alone is $\pm^1/_8$ in. Other errors are extras.

Reading Tape Measures

Four good tape measures were altered in length so that a reference scale was needed to calculate the correct lengths. The fifth tape measure was left to manufactured length. Repetitive measurements of the same distance could be made without the operator being prejudiced by his prior measurements.

The interest was in the inherent errors of determining a water surface from a good reference point, rather than skills of people. So, seven people well accustomed to using tape measures measured the distance from a good reference point to a visible water surface with the five tape measures. They were exceptionally careful. Only one person on one of three occasions achieved complete agreement among five trials. He was the person conducting the study program, and he became quite skilled in avoidance of errors. At that, he remarked that his one success was likely due to chance.

The ranges of five repeated measurements of a same distance with the tape measures varied from 0 to ½ in. As said, the zero range occurred only once and it was the culmination of much time spent in development of skills and could have been chance. The arithmetic and geometric mean ranges of errors were both about 3/16th in.

Common steel tape measures are excellent devices for carpenters. They were not designed for measuring water levels in holes.

Significance of Errors

Although most technicians read their tape measures to the nearest ¼ in., assume a technician recording to 1/8 in.

At the beginning of the last 30 min of a USPHS test, suppose the water level is actually 35-⅜ in. below mark. The technician might measure 35-¼ in., 35-⅜ in., or 35-½ in. Then, suppose the water level at the end of the 30-min period was 35-⅞ in. The technician might measure 36 in., 35-⅞ in., or 35-¾ in. Now he subtracts his measurements to convert his last 30-min readings into a percolation rate.

If both measurements are too high, too low, or correct, the technician gets ½ in., or 1 in./hr (60 min/in.). Possibly, however, he got the first measurement too low and the last too high, or the first too high and the last too low. Those combinations would yield ¾ in. or ¼ in., respectively. So, the technician walks into the local health department with a percolation rate which could be 40 min/in., 60 min/in., or 120 min/in., depending on his chance readings.

The reader is assured that foregoing thoughts deal with theoretical aspects of errors. The real errors are larger and commonplace.

Avoidance of Errors

As said, float gauges are simple, inexpensive devices designed for accurate measurements of changes of water levels. Local authorities would be well justified to require that float gauges be used in performance of percolation tests. Most gauges can easily be accurately read to 0.02 in.

CHAPTER 5

SLOW PERCOLATION RATES

The *Manual of Septic-Tank Practice* [15] notes that if soils had a percolation rate slower than 60 min/in., those soils were "unsuitable for absorption systems." That limit applied to disposal trenches. Disposal pits are advantageously deeper than trenches. Without justification, the *Manual* set 30 min/in. as a percolation rate limit for pits. One would like to believe that 60 min/in. (or 30 min/in.) really separated suitable from unsuitable soils, but it does not. Moreover in several localities, the slowest acceptable percolation rate is a political choice.

EXAMPLES OF POLITICAL CHOICES

In Florida, a soils map displays kinds of sands. With a few exceptions—marl for one and spodosols for another—Florida's sandy soils are very permeable. So when septic-tank practices in Florida were reviewed in 1977 [44], the slowest acceptable percolation rate was found to have been set at 15 min/in.

In Marin County, California, soils are not varieties of sands. They are largely varieties of heavy clay loams. In that county, there is a very liberal percolation test procedure. It yields results that must be about twice or thrice as fast as what the USPHS test would yield. And, the slowest acceptable percolation rate (at least when the writer left California three years ago) was 120 min/in.

In San Mateo County, California, soils are different again. They are far from being as permeable as Florida's sands, but they are not quite as slowly permeable as most of Marin County's soils. So, the USPHS test is used in San Mateo County, but the slowest acceptable percolation rate is 0.75 in./hr (80 min/in.).

Students of government could readily submit: one would expect to find many areas where percolation rate limits are set to accept the majority of situations, and to reject a tolerable few. In Marin County, for example, where there are many rural residents, no one acquainted with the governmental process could conceive of local authorities long surviving a decision to set a percolation rate limit which would result in refusal of the large majority of applicants for septic-tank systems. More than one person has observed that presence of septic-tank systems or public sewers is not much related to soils, rather it is a matter of population densities and wealth.

SUITABILITY OF SLOWLY PERMEABLE SOILS—TWO EXAMPLES

As a first example, San Mateo County, California, advises use of the USPHS percolation test. But whereas the USPHS notes 60 min/in. as the limit of soil acceptability, San Mateo County uses 0.75 in./hr (80 min/in.). To compensate for acceptance of slowly permeable soils, San Mateo County provides septic-tank practices improved beyond ordinary [45]. Two full-sized disposal fields are required to facilitate alternation of fields. Authorities also adhere to considerations of best assimilation by placing trenches on a same contour where possible. Doubtless through excellent practices, San Mateo County has an enviable record of successful septic-tank systems. In 1980, it was noted that 800 septic-tank systems had been constructed in accordance with improved regulations followed since 1969. No problems with those systems had been noted [46].

In Marin County, as a second example of suitability, an engineer subdivided a small property into about a dozen lots. He performed percolation tests according to local guidelines. In less than one year after residents moved into their new homes, several septic-tank systems failed. An outside consultant noted, "It would be almost impossible for a skilled technician to fail to pass the Marin County percolation-test requirement in the vast majority of the world's soils." It was that liberal. He then directed a backhoe to dig holes exposing soils in profile at the problem subdivision. USPHS percolation tests were performed. Household wastewater volumes were measured.

It was found that two neighboring couples were each wasting about 500 to 600 gpd, that being more than thrice the wastewater flows from ordinary households with only two people. Because law in California assigns responsibility of outcome of a septic-tank system to the contractor for the first year, a chance existed that each couple was deliberately testing their systems before the first year passed.

The soils of the troubled site had long ago been deliberately compacted for commercial chicken ranch buildings. Soil profiles displayed poorly permeable soils. Percolation rates were slow. One carefully performed USPHS test yielded 0.2 in./hr (300 min/in.). The disposal trench layouts did not take best advantage of the assimilative capacities of the sites, and prior liberal tests had resulted in underdesigned systems.

Despite the poor soils and very high wastewater flows, construction of additional disposal trenches yielded satisfactory systems. Most sanitarians have seen satisfactory septic-tank systems where percolation tests had failed to yield acceptable rates.

LIABILITY OF SOILS TO DAMAGE

Many soils in the United States are regarded as unsuitable for septic-tank practices. Some opinions come from unhappy experiences, and others come from belief that slowly permeable soils are not usable.

Although farmers know that tilling wet soils, especially clayish soils, destroys their permeabilities, that information has yet to find expression in septic-tank practices. On one occasion, not unusual, a system of disposal trenches was newly excavated in one of the Southeastern red soil states. Rains came and brimfilled the trenches. Later, the contractor pumped those standing waters out and finished his construction. One needs not wonder about the sad outcome.

Generally, red soils are permeable and well drained. But such soils commonly are clayish and easily damaged from digging when wet.

COMMENTS

It is not reasonable to believe that a soil with essentially no permeability would serve for a practical disposal field. As was pointed out by Laak [6], a site must be evaluated for its capacity to transmit percolating waters away from a disposal field. Such considerations of assimilative capacity are more critical in less permeable soils.

Local authorities must face the public. Until competently staffed public entities are commonplace experiences, and they are responsible for design and other aspects of septic-tank practices, local authorities can be relied upon to set a percolation test limit for administrative convenience. After all, local authorities are in most places not authorized by law to design septic-tank systems. Nevertheless, those authorities would not be unjustified in granting variances to guidelines when they judge it proper to do so. Further, improved septic-tank practices in the majority of our states would accommodate more liberal limits of soil acceptability.

The *Manual of Septic-Tank Practice* [15] was technically wrong in advising 60 min/in. as an uppermost limit of soil acceptability. A superior design approach would have better served a citizen wanting to realize a benefit from his land. Doubtless inadequate septic-tank regulations have needlessly divested many people of property rights.

CHAPTER 6

ASSESSING SOILS FOR DISPOSAL PITS

Especially in the Southwest, there are many soils situations where more permeable substrata are below less permeable soils or a restrictive stratum. There, disposal pits, sometimes several scores of feet deep, serve well. Judgments of proposed installations, however, must be made without the help of standard percolation tests. It is neither practical nor safe to perform such tests in deep excavations.

ASSESSING SOILS FOR SHALLOW DISPOSAL PITS OR DEEP TRENCHES

Where the contemplated disposal field is to be no deeper than about 12–15 ft, a backhoe can dig a hole to such depths. If the sides are not vertical, but are sloped outwards and a ramp is dug at one end, a technician can enter and work in such a hole. He can judge soil strata exposed in profile. He can perform percolation tests at the bottom of the hole, hand auger to greater depths, if desired, and he can perform percolation tests along the ramp in strata of interest through which the ramp passes. (See also Chapter 11.)

In sloping terrain, the hole should be dug along a contour so that a disposal field subsequently constructed there will not have a disturbed lower soil berm capable of conducting effluents to daylight. In flat terrain, the hole should be dug along courses a contemplated disposal trench system would take.

ASSESSING SOILS FOR DEEP DISPOSAL PITS

Where very deep disposal pits are contemplated, theoretically percolation tests in multiplicity should be performed in each stratum. But, as said, it is neither practical nor safe to do so. In the absence of techniques which could be recommended, what information is available is offered.

Existing Experience

Where a deep disposal pit is contemplated, the locations of deep strata are generally known and disposal pits are commonplace. There, site investigations could well be exploratory to find the strata known to be useful and their extent. History of existing systems could perhaps best guide decisions. Properly understood, experience is the best guide.

Full-Scale Percolation Tests and Visual Judgment of Soils

The USPHS studied disposal pits in several states [14]. Recognizing the impracticality of using standard percolation tests for guidance, efforts to correlate soil qualities to performance of disposal pits were made. Bulk and core samples were analyzed for density, texture, moisture, and porosity. Soil penetrometer tests were made on each stratum in place. What was termed a full-scale percolation test was conducted, and a system of visual judgments tried.

It was found that mass density, soil texture (proportions of sand, silt, and clay), and penetrometer values were too weakly related to performance of pits to be of practical use. Performance was judged economically to be the design area for optimum financial return. Efforts to relate performance with the full-scale infiltration test and visual ratings showed relationships met with enough success to warrant investigators noting, "These figures [reproduced herein as Figures 16 and 17] provide a reasonable means for estimating design size for seepage pits in a variety of soils."

Full-scale percolation tests were made as follows [14]:

1. Excavate a conventional seepage pit of the same diameter and depth as the operating pit within 20 feet and not closer than 10 feet to that pit.
2. Install a lining of the type used in the area being studied and place one foot of gravel in the bottom of the excavation.
3. Fill with clear water, through the use of tank truck or other suitable means, to the elevation of the ground suface.
4. Allow the water level to recede to one-half the depth of the pit or until the depth of the water is 5 feet, whichever is least. Refill and allow to drain

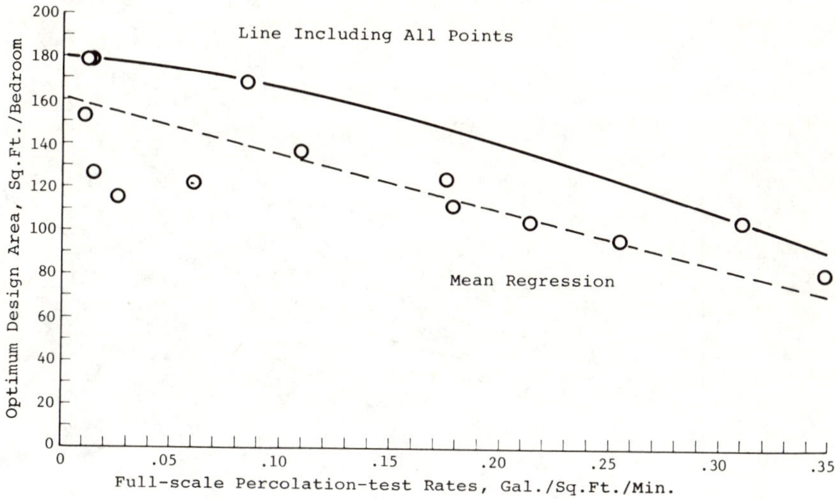

Figure 16. Optimum design areas for deep disposal pits as related to full-scale percolation test rates [14].

> overnight. On the second day refill again and time the recession of the water level to the mid-depth or the 5-foot depth, whichever is least.
>
> 5. From a plot of the rate of recession versus depth of water in the pit, the percolation rate in units of gallons per minute per square foot of wetted area is calculated over any depth of the pit desired.

Regressions are calculated by assuming one variable as independent and the other as dependent. Not knowing which gave better data—the full scale percolation test or the optimum design area—two regressions were calculated by interchanging independent and dependent variables. Slopes were only slightly different, so a mean regression line was drawn between the two calculated regression lines, and it was entered in Figure 16.

In disagreement with conclusions of the investigators of the USPHS who advised the use of their regression line, it is pointed out that a regression line has, and is supposed to have, about as many data points above the line as below it. Use of the regression line would therefore result in underdesign of about half of the systems. Therefore an envelope curve, a line including all points, was drawn in Figure 16. That line would constitute a better design criterion, assuming that the data adequately delineated it.

Visual judgments were described as follows [14]:

> Each soil stratum encountered during excavation of the pit was rated on a scale of one to four. The rating was based primarily on visual evaluations of texture, structure, density, and degree of cementation or compaction. A description of the four visual rating classes is as follows:

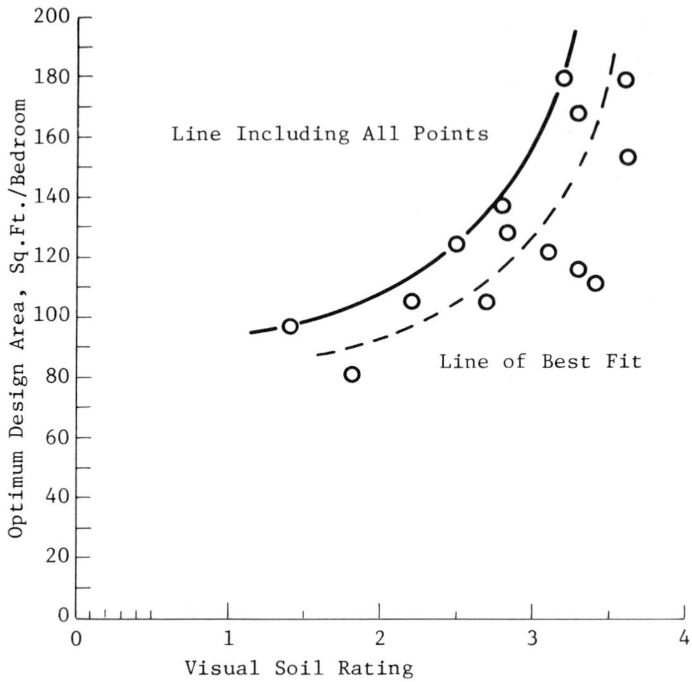

Figure 17. Optimum design areas for deep disposal pits as related to visual soil ratings [14].

1. Clean gravel or sand, very loose.
2. Clean gravel or sand, loose to firm;
   Sandy loam, friable.
3. Clean gravel or sand, firm to dense;
   Sandy loam, firm to dense;
   Loam to sandy clay, friable.
4. Silty clay or heavier, dense to very dense.

Individual stratum ratings were weighted for thickness of stratum in determining the average rating for the entire pit or over any depth of the pit desired.

Results presented in Figure 17 had been fitted with a regression line by USPHS investigators: optimum design area = 35 + 33 visual soil rating. Study of the data, however, does not support belief that a straight line relationship was delineated. Human senses (visual judgments) are known not to respond to stimuli in direct proportions. Rather, human senses offer curvilinear responses. In line with those thoughts, data were grouped, averaged, and fitted with the curve given in Figure 17. As before, it was reasoned that a line including all points would conform to best practice, and that line was also given.

Foregoing efforts were made to present visual soil ratings in a best light, but note that the data points in Figure 17 are quite scattered. Further, visual judgments were made by persons especially skilled in both soil sciences and the particular field investigations. It would not seem likely that the USPHS scheme of visual judgments would constitute a useful general method.

During studies of a subdivision, a local technician was employed. He was thoroughly experienced in laboratory soils analyses as well as fieldwork of performing percolation tests and visually judging soils in terms of soil types, scaled 1 through 5, most to least permeable, as given in the *Uniform Plumbing Code* [47]. He was the most highly respected technician in that rather knowledgeable county.

During studies of exploratory excavations too deep for standard percolation tests, the skilled local technician collected excavated soil samples of rather sandy soils which he later subjected to sieve analyses. Visually, he assigned *UPC* soil type designations in the field, which judgments he later confirmed with laboratory analyses. The writer collected samples at the same time in the field. He measured permeabilities of those samples, replicated at locations, to obtain mean measurements of Darcy's coefficient of permeability. The results of measurements (less graphs of textural analyses) are presented in Table VII. There it can be seen that very poor agreement was had between experienced visual judgments of soils and objective measurements of permeability. For reference convenience, coefficients of permeability

Table VII. Visual Judgment Compared to Measurements of Permeability[a]

Sample No.	Coefficient of Permeability, (ft/day)	Computed Percolation Rate (in./hr)	Computed Percolation Rate (min/in.)	Visual Judgment, UPC Soil Type
1	170	191	0.31	1
2	85	121	0.50	1
3	23	52	1.1	2
4	11	33	1.8	5
5	6.8	24	2.5	3
6	1.2	7.6	7.9	5
7	1.1	7.5	8.0	2
8	1.1	7.5	8.0	3
9	0.40	3.8	16	3

[a]Note: Percolation rates estimated from measurements from a falling-head permeameter.

have been translated into percolation test rates by formula (Chapter 3, page 30).

The writer is convinced that it is a rare individual who can judge soils visually with more than modest accuracy. From the practical point of view, visual judgment does not constitute a technique suitable for general practices, except in judgment of extremes, i.e., very good or very bad soils.

A Very Conservative Approach

There is another suggested approach to the problem of assessing a site for deep disposal pits. An exploratory hole could be dug with a bucket auger of small diameter. Samples of soils could be visually judged for minimal permeability. Then if deep strata seem permeable, even slowly so, disposal pits could be designed to meet worst-case situations. Discussions of design criteria (Chapter 7) will present the Most Conservative Criterion, which could be used. From a practical point of view, it might be less expensive to overdesign a disposal pit system than to undergo extensive explorations of deep soil strata.

CHAPTER 7

SIZING A DISPOSAL FIELD

During lectures and seminars, the various methods of sizing a disposal field are usually presented. Faced with a variety of choices and feeling somewhat at a loss, someone always asks, "Dr. Winneberger, in your practice how do you size a field?" The answer is, "I see how much my client can afford." When the laughing ceases, the answer is elaborated to compromise ignorance, theory, and practicality. A bigger system will likely give years of performance, far beyond simple proportions of costs.

There are three attitudes towards sizing a disposal field. Firstly, there is the belief that a disposal field must be sized large enough to serve at all—something like a bridge must be strong enough to bear traffic and weather. Secondly, there is a belief that all disposal fields have distinct lifespans and should, therefore, be considered temporary devices. Thirdly, and rarer, there is the belief that disposal fields could serve indefinitely, and it is in this belief that the writer joins his friend, Professor Rein Laak, a leading proponent [6], in convictions.

LOADING A SOIL

The design size of a disposal field relates to its inundation with effluents from a septic tank. A loading may be viewed in terms of volume, time inundated, and strength of the effluent.

Volumetric Loading

Most commonly used criteria for loadings relate the surface area of a disposal field to volumes of effluents. The surface area is the total area which could be inundated rather than some area actually to be inundated from the onset. Volumes of effluents are sometimes expressed in terms of gallons per square foot per day (gal/ft^2/day). More commonly, loadings are related to bedrooms a house is to have, people to be served, or in some other way gal/ft^2/day has been translated into maximum water usage expected to reach a disposal system.

Time Loading

Research disclosed loadings to be time related. Early in studies, it was learned that the rate of soil clogging was essentially independent of volumes of effluents infiltrated [41,48,49]. Instead, the length of time soil rested after infiltrating effluents and before receiving more, had a strong relationship to clogging. Volumes, nevertheless, would affect the length of time a particular increment of soil was inundated before draining.

Waste Strength Loading

The soils of a disposal field contain myriads of microbes, and household wastewaters provide foodstuffs for those microbes. One might expect that a greater supply of foodstuffs, the strength of the wastes, would result in greater or lesser numbers of microbes. And because presence of microbes and their by-products are associated with reduced soil permeability, one might believe that wastewater strength would be related to size of a disposal field. Nevertheless, that consideration constitutes a moot issue.

In 1960, investigators at the University of California, [48] compared septic-tank effluents (anaerobic) with effluents from an extended aeration device (commonly termed aerobics) as to soil clogging capabilities. It was found that in a coarse soil, aerobic effluents produced an earlier but less intense clogging than did anaerobic effluents. In a fine soil, however, the reverse was true. Nevertheless, from a practical point of view, the two effluents were regarded as equal.

In 1966, Laak [50,51] related the amount by which a standard disposal field could be reduced to a function of improved wastewater pretreatment, specifically aerobics. Laak focused on concentrations of 5-day biological oxygen demand (BOD_5) and total suspended solids (TSS). Laak noted:

$$\text{Adjusted Area Required} = \text{Area Required for Standard Septic-tank Pretreatment}$$
$$[(BOD_5 + TSS) \div 250]^{1/3}$$

Later, Otis et al. at the University of Wisconsin studied aerobics and septic tanks. In 1977, Otis summarized [52]: ". . typical aerobically treated wastes do not result in less clogging than septic-tank effluent in fine textured soils where the difference would be most desirable. . . . The story is different with sand filters."

Several proprietary devices purport to pretreat wastewaters better than septic tanks do. Properly maintained, some doubtless do. As a practical matter, however, most such devices do not get the maintenance they need. At least two surveys disclosed the almost complete failure of homeowners to maintain sewage-treatment units. Even when a demonstration project for public management of aerobics was surveyed, an authoritative report [53] noted, "As a whole the project, at this writing, must be regarded as a failure." Aerobics can be maintained, but the writer is aware of only one story of success, that is, maintenance provided by a small private business [46].

Notwithstanding maintenance problems and assuming Laak's formula applies, when one calculates the cube root of the BOD_5 and TSS from the usual aerobic, the reduction does not amount to much. Pretreatment costs, especially ongoing energy costs for machinery, exceed savings of a reduced disposal field.

LOADING AND CLOGGING SOILS WITH FRESH WATERS AND WASTEWATERS

Fresh Water Clogging

When a soil is inundated with fresh water, permeability of that soil changes. Winterer [54,55], observed water spreading on soil cores. He found the infiltration rates decreased rapidly for about 10 days (Phase I), then increased for about 25 days (Phase II), and thereafter decreased slowly to reach a relatively low level after several months (Phase III). A typical curve of long-term soil acceptance of fresh water is presented in Figure 18. With numerical variations, such a curve has been observed of essentially all inundated soils whether tested in a laboratory or in the field.

It has been generally agreed that Phase I results from slaking of soil, that being a result of a change in both the affinity of the internal soil surfaces for water and the cohesive forces holding the system together [56,57].

See Figure 18 and note that the short-term percolation test occurs during early stages of Phase I. It would appear that efforts to measure stabilized rates would be impossible during the very short-term percolation test. Actually, what is referred to as a stabilized rate is a temporary

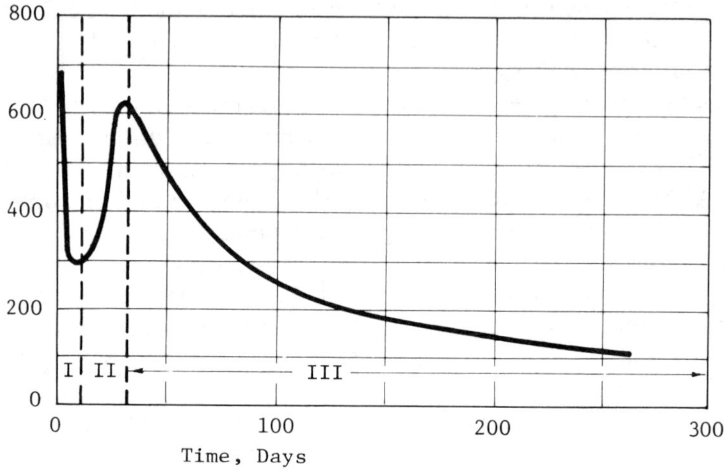

Figure 18. Typical percolation rate curve of prolonged fresh water spreading on a soil core occurring in three phases [54].

achievement of satisfaction of capillary forces and some quiescence of damage to soil surfaces from digging and adding water.

Phase II, a recovery in permeability, results from removal of air entrapped in soil pores by its solution into percolating waters [58]. Phase II has implications to the percolation of water away from the immediate vicinity of a disposal field. Although a very small soil pore may have a greater capillary attraction for water than a larger pore would, solution of air in the larger pore would draw percolating waters into it. Once within both smaller and larger pores, water would percolate preferentially through the larger pores because of less resistance to flow.

Phase III, the long-term decline in permeability of a soil, has been shown to result from biological activities [59]. Under sterile conditions, Phase III does not occur, and the maximum of Phase II is maintained.

During Phase III, the greatest resistance to percolation occurs in the upper 3–6 in. of a soil core, wherein organic foodstuffs for microorganisms are located [60].

Wastewater Clogging

Whereas soils undergo a 3-phase permeability change in response to long-term inundation with fresh waters, those phases compress when wastewaters—sewages—are spread. The injection of high levels of microbial populations with their foodstuffs brings on the third-phase clogging rapidly. A curve depicting typical loss of permeability during wastewater spreading is presented in Figure 19.

Whereas fresh water inundation results in a reduction of permeability in a 3- to 6-in.-thick portion of a soil, wastewater spreading results in clogging of the wastewater-soil interface. The clogged interface is generally only an inch or so thick. Soils below the clogged surface (also biocrust, biological mat, etc.) remain relatively unaffected.

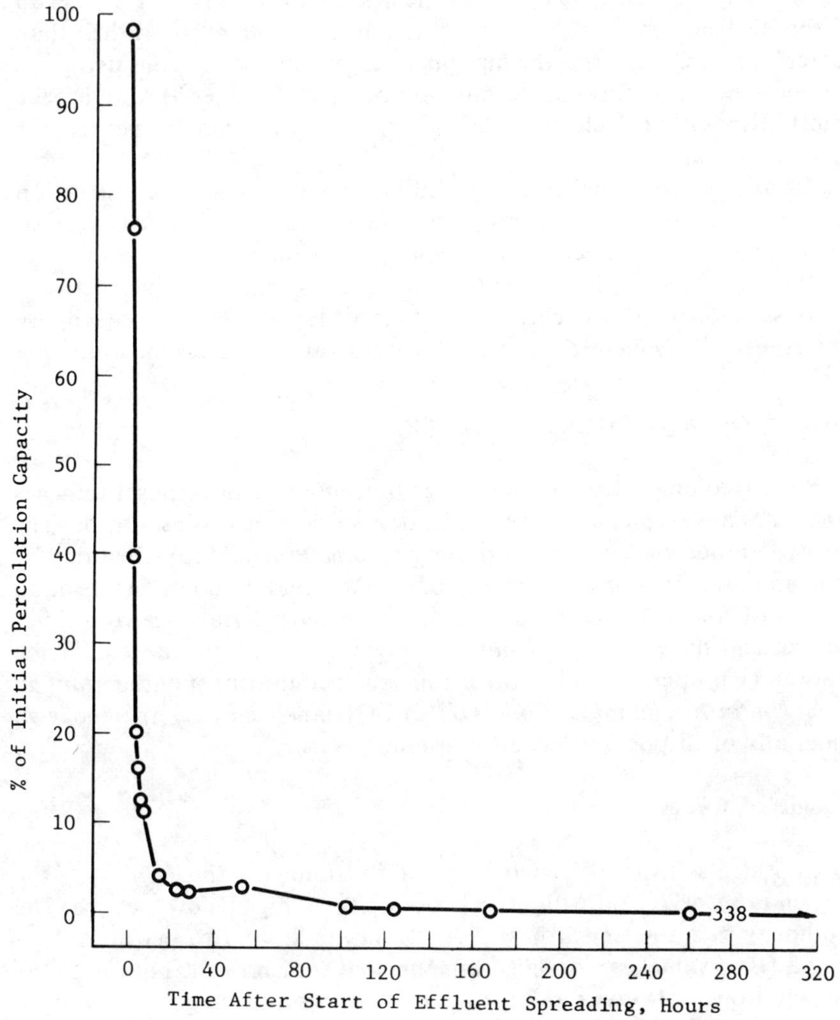

Figure 19. Rates of percolation of septic-tank effluent through Oakley sand.

Practical Implications

Clogging being superficial has implications to design of disposal fields [61]. Firstly, maximizing the exposure of soil surfaces in a disposal field to effluent inundations is suggested. Ideally, a disposal field would consist of a huge crack in the ground. Practically, the designer can reduce the ratio of gravel-filled volume to the soil surfaces which could be inundated.

Secondly, clogging being largely restricted to a thin surface layer in disposal fields means that infiltration into a soil mass, rather than percolation through it, is the limiting phenomenon. Thus, considerations of construction needs and assimilation benefits (Chapter 9) would space most disposal field elements farther apart than might be needed for percolation alone.

Thirdly, where trenches are spaced far enough apart for construction convenience, additional disposal field elements might be constructed between existing elements where space on a lot is otherwise in short supply. In other words, there is nothing wrong with soils between distressed disposal trenches. Still, the reader is not advised to use interspaces if avoidable because of considerations of assimilation.

SOIL INFILTRATION SURFACES

Rather commonly, codes recognize bottom areas of disposal trenches and sidewalls of pits as active infiltration surfaces in disposal fields. The revised edition of the *Manual of Septic-Tank Practice* [15] benefitted by ongoing investigations of septic-tank systems, and it noted, "Absorption area is figured as trench bottom area, and includes a statistical allowance for vertical sidewall area." Unfortunately, the *Manual* failed to make the allowance properly, and bottom areas remained the thrust of the *Manual.* The *Uniform Plumbing Code* (*UPC*) [47] failed entirely to recognize sidewalls of disposal trenches as useful.

Illogic of Codes

Consider a disposal trench in a county following the *UPC*. Say the trench is 3 ft wide and with vertical sidewalls 1 ft high below the pipe. The authority says the trench has 3 ft^2 of useful bottom area per lineal foot and 2 ft^2 of valueless sidewalls. Imagine that somehow the bottom of the trench drops 19 ft lower and the trench becomes a rectangular pit with 20 ft of total sidewall height. Now the authority adds the theretofore valueless 2 ft^2 of sidewalls to the additional 38 ft^2 (2 sidewalls). He credits the pit with 40 ft^2 of infiltration surfaces. He now discredits the theretofore accepted 3 ft^2 per lineal foot of bottom surfaces just because they are lower and the device changes names. The ruling is, of course, without justification.

Sidewalls vs Bottoms—A Moot Issue

Investigators of the USPHS, having studied "Permeability Conditions in Operating Tile Fields" (disposal trenches) noted [33], "The evidence that lateral movement of water is a factor in effluent absorption by this soil suggests the possible use of narrower but deeper trenches for more efficient tile line systems."

Studies at the University of California produced data causing the investigators to regard sidewalls as more useful than bottom surfaces for infiltrating wastewaters [62]. Direct measurements displayed a marked tendency for wastewaters to infiltrate sidewalls in preference to bottom, horizontal, soil surfaces. Later those investigators studied periods of cessation of spreading wastewaters onto soils and consequent resting of those soils. Whereas sidewalls recovered much of their original infiltrative capacities, bottom surfaces did not [49].

Studies at the University of Wisconsin convinced those investigators that infiltration occurred mostly in bottom surfaces of disposal fields. But after a time, those investigators noted [63], "In some soils the sidewall may become the more significant infiltrative surface as clogging continues." And, "In an unclogged absorption system, the potential gradient is lower for the sidewall area than the bottom area because the gravitational potential is zero. As the clogged zone develops, the matric potential of the bottom area may be reduced to where the sum of the gravitational and matric potential is less than the matric potential of the sidewall area. The sidewall then becomes the dominating infiltrative surface." Still, they went on to note, "Maximizing the infiltrative area through consideration of sidewalls as a reserve capacity is recommended but in the Midwest, the bottom area should be sized to absorb the entire estimated daily flow."

Laak [6] pointed out that clogging was an expected and not deleterious event. It offered treatment of wastewaters as they infiltrated the "biocrust" (also clogging mat, biological mat, and other terms for the thin zone of reduced permeability at a wastewater-soil interface). Laak noted, "The sidewalls and bottom areas of soil surfaces could be considered as effective infiltration surfaces."

Human psychology has most of us following our own participatory experiences in preference to those of others. Only when outside evidence is overwhelming will we alter our position, first to equivocations, then perhaps to acceptance (or retirement). So until the matter of bottom vs sidewalls yields to knowledge compromising all data, the issue remains moot and authorities of the land will have a confusion to live with. Meanwhile, there is a practical direction a designer of a disposal field may take.

A Practical Attitude

Gravel is a major cost in construction of disposal fields. In some places, gravel is so expensive that alternative backfillings of rubble, gravel made of fired clay, pieces of plastic pipe, and such are used. It would be to advantage to use design criteria which require the least gravel for the soil surface area obtainable. And regardless of personal convictions of sidewalls vs bottom areas, local authorities rule design. It is the designer's task to obtain the most efficient system, that is, the most active infiltration surfaces for the cost.

Figure 20 is self-explanatory. It offers rationale for considering "narrow and deep" as design attitudes leading to efficiency. Although the *Manual of Septic-Tank Practice* notes, "Absorption area is figured as trench bottom area and includes a statistical allowance for sidewall area," it fails to do so. Consequently, unless local authority allows proper arithmetic for accounting for sidewalls, following the *Manual* will sometimes lead to deep trenches with greater sidewall area than the equivalent shallower system. Nevertheless, within the faulty framework of the *Manual*, a designer can trade off advantages and disadvantages and still find "narrow and deep" attitudes best economy.

USPHS FORMULA FOR SIZING A DISPOSAL FIELD

The USPHS studied ordinary septic-tank systems which had not been intended to serve indefinitely—they had lifespans. Field data yielded an empirical general formula for designing systems expected to have a chosen probable lifespan:

$$A = Q/L\ (K^{-1} + t)$$

The quantity A is the bottom area of disposal beds and sidewalls of pits. The investigators [20] were convinced, "A given increment of the total aborptive capacity will be exhausted each year . . . ." Thus, t is the years of service expected for a given area, A. The quantity k is a soil parameter representing the rate at which soil becomes clogged to a nonserviceable state. L is a soil parameter representing the volume of liquid that the soil can ultimately infiltrate. Q is a loading factor which includes several

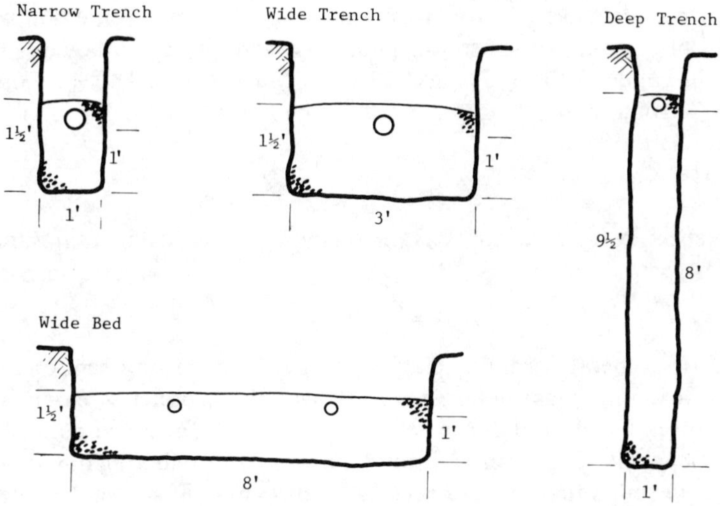

ESTIMATES PER LINEAL FOOT

Geometry	Gravel	Bottom Surface	Functional Surface	BS/G, ft^2/ft^3	FS/G, ft^2/ft^3
Narrow Trench	1.5 ft^3	1 ft^2	3 ft^2	0.67	2.
Wide Trench	4.5 ft^3	3 ft^2	5 ft^2	0.67	1.11
Wide Bed	12 ft^3	8 ft^2	10 ft^2	0.67	0.83
Deep Trench	9.5 ft^3	1 ft^2	17 ft^2	0.10	1.79

Conclusions: If local authority credits bottom surfaces of trenches only, as infiltrative surfaces, the narrow trench offers the same credited surfaces per unit of gravel, BS/G (Bottom Surface/Gravel) as does the wide trench. Still, the functional surfaces in the narrow trench per unit of gravel, FS/G (Functional Surface/Gravel) exceed that ratio in the wide trench. Local authority would likely credit the bottom of a bed and the sidewalls of a deep trench as infiltration surfaces. The functional surfaces of the narrow, deep trench per unit of gravel (FS/G) exceed that ratio of the wide bed.

Figure 20. Arguments for deep and narrow disposal field elements, rather than shallow and wide.

subfactors, such as sewage characteristics, the amount of treatment the sewage receives in the total system before the septic-tank effluent comes into contact with soil, the number of persons using the system, and the total volume of sewage discharged to the system.

Through considerable field studies of existing septic-tank systems in a locale, constructed in a specific soil type, and involved mathematics, a specific formula could be derived from the general formula. For example, existing septic tanks with disposal pits constructed in accordance with practices in vogue in the 1950s and 1960s in Maryvale Parks, Maricopa County, Arizona, could be designed for a chosen probable lifespan by substitution in:

$$A = 37.4\ t$$

Another specific formula for East Broadway Additions in that same county [14] was

$$A = 18.2\ t$$

The USPHS general formula relating lifespans of existing septic-tank systems to existing construction practices was the product of extensive studies. More of such studies are needed to render the formula applicable to a specific locale. The formula is totally useless for predicting outcomes of either new soils situations or new construction practices. If one wanted a system with an exceptionally long lifespan, the formula should not be used. It was only valid within the lifespan of systems observed. The USPHS formula neither found, nor could find, practical application.

The USPHS formula experience constituted a major effort to develop design criteria for septic-tank systems with expected, probable lifespans. The hapless result is an excellent argument for using criteria to design disposal fields expected to have indefinite lifespans.

AN IMPORTANT EFFECT OF SIZE

A larger disposal field might give years of performance beyond simple proportion to costs. Not all soils respond alike, but the point is made as follows.

During USPHS investigations [64,65] of septic-tank systems in Decatur and Sequoia soils in Knox County, Tennessee, the specific formula was derived:

$$A = 263 + 26\ t$$

Thus a residential disposal bed of 400 ft^2 would serve a mean of 5 years. If the bed offered 600 ft^2 of bottom surfaces, it would serve about 12 to 13 years. And if the bed offered 800 ft^2, twice the 400 ft^2 that lasted 5 years, that disposal bed would serve a mean of 20 years, far beyond direct proportion.

The data refute a comon belief that a disposal field is comparable to, say, a cable, which must be strong enough to support a certain load, There, purchase of a much stronger cable would likely constitute waste. Disposal fields are not comparable. Somewhat larger fields are generally better, at least beyond most local practices.

RYON'S CRITERIA OF VOLUMETRIC LOADING

With changes in expected wastewater flows from a household, or with loading rates expressed in terms of bedrooms (a very poor index of expected wastewater volumes) or the like, tables relating a percolation test to design size of a disposal field are varieties of the work of Henry Ryon. They all assume the design size of a disposal field to be related to volumetric loading rates.

When Henry Ryon devised his percolation test, he visited troublesome septic-tank installations. At each, he performed his test in unaffected soil. He determined the size of the troublesome disposal field, and he estimated volumes of septic-tank effluents disposed to the field. From those data, he calculated hydraulic loading rates onto soil surfaces of bottoms of disposal fields and all surfaces of cesspools, except the bottom of the first cesspool connected in tandem with others.

Ryon was apparently aware that the disposal fields he studied had lifespans. Believing that 20 years was an adequate lifespan, he excluded older systems from his studies. He apparently studied both leaching cesspools and disposal pits and referred to both as leaching cesspools. He likely accepted whatever dimensions leaching cesspools happened to have, but he limited his studies of disposal trenches (which he termed tile fields) to those which were 1 ft wide. The height of the gravel filling surrounding a centrally located tile pipe (perforated pipe nowadays), cannot be accurately discerned from literature. Ryon's drawings [22,66] noted gravel fillings both 8 and 10 in. high. Likely, field practices favored 8 in., the earlier reported of the two values.

After 27 disposal fields and 21 cesspools had been included in Ryon's studies, he plotted loading rates of septic-tank effluents against percolation test rates. Then assuming that the troubles with the systems were caused by excessive volumetric loadings, Ryon drew an envelope curve under most of the data points below which he assumed loading rates were proper for a 20-year lifespan. That curve, presented in Figure 21, thenceforth became the foundation of septic-tank practices the world over, filling a vacuum of ignorance.

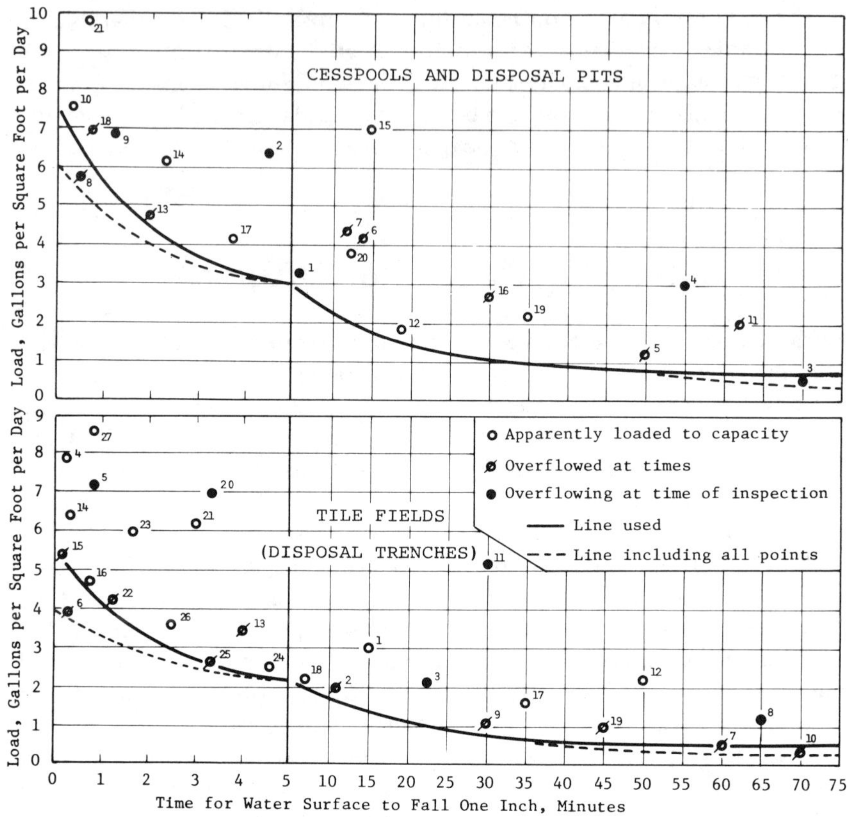

Figure 21. Permissible loadings for subsurface irrigation systems for various percolation test rates (after Ryon in Ref. [66]).

Disposal Trenches

Using Ryon's recommended tables [22,67] and making arithmetical deductions led to data presented in Table VIII. The data in the table approximately correspond to Ryon's graph, but Ryon apparently used data other than his envelope curve for assembly of his tables.

In Table VIII, Ryon's loading rates have been translated into what those rates would have been if Ryon had considered surfaces other than bottoms of trenches as useful infiltrative surfaces. (As mentioned, the expression Functional Area herein assigns equal value to bottom and sidewall surfaces, although the writer regards sidewalls superior.)

In Table VIII, the top surfaces of disposal trenches were not included in

Table VIII. Ryon's Percolation Rates Related to Various Parameters of Loading Rates of Septic-Tank Effluents onto soils of Disposal Trenches[a]

Coefficient of Permeability (ft/min)	Percolation Rate (min/in.)	Loading Rate (gal/ft²/day) Bottom Area Ryon	Bottom Area Federick	Sidewall Area	Functional Area	LTAR
	1/6	5.3	12.	4.0	2.3	
	1/4	5.0	10.	3.8	2.1	
	1/2	4.7	7.1	3.5	2.0	
	3/4	4.4	5.8	3.3	1.9	
2.0×10^{-2}	1	4.0	5.0	3.0	1.7	0.80
6.7×10^{-3}	2	3.2	3.5	2.4	1.4	0.58
3.6×10^{-3}	3	2.8	2.9	2.1	1.2	0.51
2.3×10^{-3}	4	2.4	2.5	1.8	1.0	0.46
1.6×10^{-3}	5	2.2	2.2	1.7	0.95	0.44
5.5×10^{-4}	10	1.7	1.6	1.2	0.72	0.37
2.9×10^{-4}	15	1.3	1.3	1.0	0.57	0.34
1.0×10^{-4}	30	0.84	0.91	0.63	0.36	0.30
5.3×10^{-5}	45	0.62	0.75	0.46	0.26	0.28
3.4×10^{-5}	60	0.48	0.65	0.36	0.21	0.27
2.4×10^{-5}	>75	0.44	0.58	0.33	0.19	0.26
1.2×10^{-5}	120	0.23	0.46	0.17	0.10	0.24
9.6×10^{-7}	600	0.08	0.20	0.060	0.03	0.20

[a]Ryon's interpretation of his data [22]. Federick's formula [67]. Sidewall Area assumed Ryon's loading rates and his drawing of trenches 1 ft wide and with 8 in. of sidewall height [22]. Functional Area assumed both bottom and sidewall areas equal, and has been related to Ryon's loading rates. Long-Term Acceptance Rates (LTAR) were calculated on the basis of k, Darcy's coefficient of permeability [6]; k was estimated by a correlation formula to percolation rates [29]; ft/min × 0.51 = cm/sec; and, LTAR applies to Functional Area.

estimates because many trenches nowadays are not designed to be inundated much above distribution pipes. Nevertheless, the trenches used in Ryon's day had the geometry of shallow subsurface drains. It is entirely possible that top surfaces of the troubled systems Ryon studied were serving to infiltrate surcharged septic-tank effluents, as well as sidewalls and bottoms.

Cesspools and Disposal Pits

During the last few decades, cesspools have found less favor than gravel-filled disposal fields, separate of the septic tank itself. Thus, Ryon's studies of cesspools have not been widely used.

Ryon presented a table giving loading rates of leaching cesspools (also disposal pits) with corresponding percolation rates. Notwithstanding loading rates of that table being about 69% of corresponding rates for disposal trenches, Ryon stated,

> In general, the number of square feet in the floors and in the walls below the flow line of the cesspools, measured on the outside of the walls, should be equal to about 75% of the number of lineal feet of tile required in subsurface irrigation fields.

By lineal feet, Ryon referred to trenches 1 ft wide and offering 1 ft^2 of bottom surface per lineal foot of tile (distribution pipe). Ryon noted,

> When no settling tank precedes the cesspools the bottom of the first one will clog rapidly and cannot be considered as effective absorption area.

Thus, Ryon regarded as useful bottom areas of cesspools beyond the first connected in tandem, and bottoms of all disposal pits.

FEDERICK'S FORMULA

Henry Ryon performed his studies to solve problems he and his colleagues faced in their duties of administrating septic-tank practices in New York State. Ryon described his work in a private publication, as it pertained to his tasks. After Ryon's criteria spread to almost universal acceptance, Joseph Federick, Ryon's colleague, published Ryon's work [66,67].

Federick also added his efforts by committing Ryon's schedule of septic-tank effluent loading of a disposal field as related to percolation test rates to approximate formulation. Federick plotted Ryon's data on log-log paper and drew a straight line of best fit. His formula for disposal trenches was:

$$Q = 5/\sqrt{t}$$

where Q is the loading rate in gal/ft^2/day and t is Ryon's percolation rate in min/in. He also provided:

$$Q = 7/\sqrt{t}$$

for leaching cesspools.

Although log-log paper tends to straighten out just about all data, Ryon's data still had a bit of a curve to them. Consequently, Federick's formula yields data best matching Ryon's in the mid-range percolation rates. As shown in Table VIII, Federick's formula for disposal trenches yields higher loading rates for percolation rates slower than about 45 min/in. and faster than about 4 min/in.

The "Joint Committee" authoring the *Manual of Septic-Tank Practice* used Federick's formula for advising designers of institutional disposal fields. Those tables will not arithmetically match tables given for residential designs. Without explanations of presentation of at least two systems of advice, what should be a self-consistent document is not [68].

USPHS *MANUAL OF SEPTIC-TANK PRACTICE*

Ryon's hydraulic loading rates could not be administered in sizing systems for residences, so he assumed 80 gal per capita per day (gpcpd) to be on the high side of the usual 50 gpcpd and related loading rates to a per-person basis.

Basis of Loading Disposal Trenches

In anticipation of writing the *Manual of Septic-Tank Practice*, investigators of the USPHS sought design criteria more easily administered than Ryon's per-person tables. Whereas a builder wanting as small a septic-tank system as allowable might quibble about the number of persons a house was to hold, the number of bedrooms is known. Studies of the relationships of occupancy to water usage and bedrooms per dwelling were made [38,69].

Table IX was assembled from a study of 296 dwellings. The investigators noticed that 68.2% of the dwellings had less than two persons per bedroom and only 8.5 had more. While perusing the table, however,

Table IX. Relationship of Occupancy to Bedrooms/Dwelling (After the USPHS)

Bedrooms Per Dwelling	Persons Per Dwelling										
	1	2	3	4	5	6	7	8	9	Total	%
1	1	9	1							11	3.7
2	2	32	64	51	17	2	1	1		170	57.4
3	1	2	13	42	14	9	2			83	28.1
4		1	2	5	6	5	1		1	21	7.1
5			1		1	2	3	1		8	2.7
6				1		1				3	1.0
Total	4	44	81	99	38	19	8	2	1	296	
%	1.4	14.9	27.4	33.4	12.8	6.4	2.7	0.7	0.3		100

one will find support for his experience. People who can afford larger homes have spare bedrooms, and large families often fit into the less large homes. When the same data were presented as in Table X, they clearly showed a strong tendency for the average number of persons per bedroom to decrease as the number of bedrooms increased.

The USPHS further showed that as the number of occupants in a house increased, the average water consumption per person decreased. For example, one resident per house used a mean of 50 gpcpd. In progression, usage dropped to 23 gpcpd with nine residents per house.

Despite knowledge that one household was far different from another, the *Manual of Septic-Tank Practice* had hyraulic loading rates expressed in per-bedroom values, without regard to house size or occupancy.

USPHS investigators provided a table relating "the original Ryon allowable sewage loading data," to adjustments for increased water usage expected from garbage grinders and automatic-sequence clothes washing machines. Those increases at about 20% and 40%, respectively, gave rise to the familiar table in the *Manual* used almost nationwide for "Absorption-area requirements for individual residences." The original Ryon data in gal/ft^2/day multiplied by corresponding USPHS tabulated loading in ft^2/bedroom, without garbage grinder or automatic washing machine, should yield the gal/bedroom value the USPHS assumed in estimates. Perchance the USPHS investigators used Federick's formula, also presented in the *Manual*; Federick's loading rates were also considered in deriving Table XI.

As shown, data in Table XI would imply that the USPHS used a variable value for estimated wastewater volumes in formulating the table of loading rates presented in the *Manual of Septic-Tank Practice*. In deriving Table XI, it can only be known that one or more of the "constants" were not kept constant—that is, wastewater volume increases from garbage grinders, or automatic washing machines, or the estimated

Table X. Number of Persons per Bedroom [38][a]

Bedrooms Per House	Average Number of Persons Per Bedroom	Maximum Persons Per House
1	2.00	
2	1.69	5.16
3	1.41	6.12
4	1.24	8.44
5	1.22	
6	0.83	

[a]Maximum Persons Per House are from approximate estimates of the 95th percentile—rates of accumulation that will not be exceeded in about 95 out of 100 cases. Omissions are where data were too limited for values to be estimated.

Table XI. USPHS Implied Relationship Between Percolation Rates and Wastewater Volumes [a]

Ryon's Percolation Rate (min/in.)	Loading Rates (gal/ft²/day)		Loading Rates (ft²/bedroom)	USPHS Implied Wastewater Volumes (gal/bedroom/day)	
	Ryon	Federick	USPHS	Ryon	Federick
2	3.2	3.5	50	160	175
3	2.8	2.9	60	170	175
4	2.4	2.5	70	170	175
5	2.2	2.2	75	165	165
10	1.7	1.6	100	170	160
15	1.3	1.3	115	150	150
30	0.84	0.91	150	125	135
45	0.62	0.75	180	110	135
60	0.48	0.65	200	95	130

[a] USPHS loading rates from [38]. Ryon's and Federick's loading rates from Table VII. USPHS implied wastewater volumes are the product of gal/ft²/day × ft²/bedroom.

wastewater usage of persons, or assumed persons per bedroom. It can only be known that arithmetical errors were a part of translating Ryon's volumetric loading rates from gal/ft²/day into ft²/bedroom values presented in the *Manual.*

Attempt to Corroborate Ryon's Study

When investigations of septic-tank systems were begun in the mid-1940s, the work of Henry Ryon constituted about all that was known. Thus investigators set about gathering more data to confirm Ryon's graph for disposal trenches [33]. As is an inclination of government-based research, conclusions were tacitly (sometimes outright) understood at the onset. So, when the data were graphed and compared to Ryon's curve, it was concluded that the data corroborated Ryon's findings. They did not.

The USPHS described 27 trouble-free disposal-trench systems, 9 systems with a history of continuous difficulties, and 9 systems with a history of occasional seepage. Inasmuch as the average system was about 5½ years old, troublefree systems had several years in which to become troublesome before reaching the 20 years Ryon believed to be a proper lifespan.

When people have occasional trouble with their septic-tank system, they become conscious of water-use habits and become secretive. In at least one of our states in the years of the USPHS investigations, a troublesome septic-tank system was handled as a criminal matter. A good chance existed troubles were more frequent than reported to USPHS canvassers. And, even occasional seepage would be regarded as troublesome by most people, so all data from the troubled systems were combined and graphed.

In Figure 22, it can be seen that a student assigned the task of drawing an envelope curve under troubled systems hydraulically overloaded would not have drawn Ryon's curve. Rather, a straight line would have been drawn running from percolation rate 4.2 min/in. at 0.49 gal/ft²/day to percolation rate 111 min/in. at 0.254 gal/ft²/day (latter point off-scale). As said, the trouble-free systems could be ignored because their short lifespans of trouble-free performance could yet be terminated before the 20-year criterion. Their existence could neither repudiate nor delineate Ryon's curve.

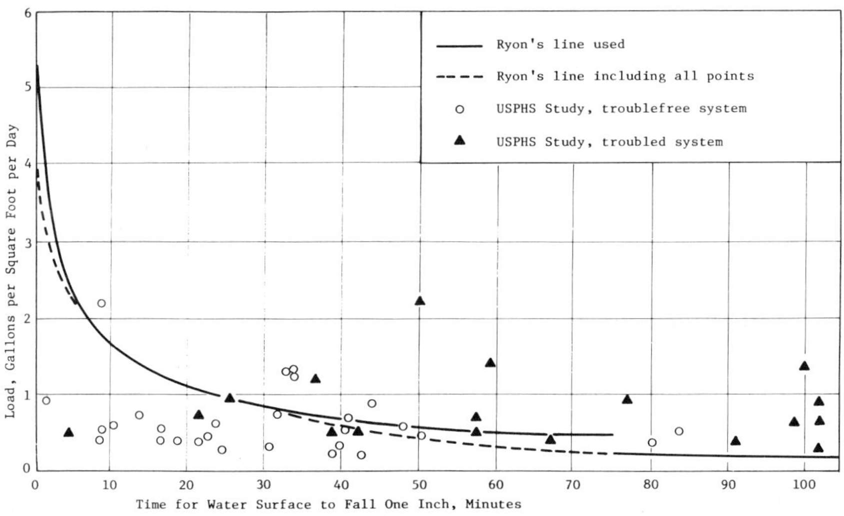

Figure 22. Relationship of disposal field loading rates to USPHS percolation test rates [33].

Regardless of the USPHS data failing to corroborate Ryon's curve, they could not be taken as discrediting evidence either. It should be borne in mind that the USPHS investigators used the USPHS percolation test, and Figures 21 and 22 are not really comparable. The USPHS test and Ryon's test are very, very different.

Commentary

It is common knowledge and a matter of history that individuals may excell, but committees with authority dispersed among members never do. Group efforts are sometimes successful with centralized authority. Despite that knowledge, our nation has a disposition towards committees and certainly an inclination to accept rulings of governmental committees, even rulings in contradiction of scientific fact.

The USPHS has, without apparent reason:

1. excluded bottoms of disposal pits in estimates, whereas Ryon included them
2. regarded infiltrative sidewall surfaces of disposal pits as equal to bottoms of trenches, although Ryon advised a 75% reduction for pits
3. changed Ryon's percolation test basically and liberalized it in the case of the less permeable and most critical soil situations
4. provided a table of wastewater loading rates arithmetically at variance with Ryon's data

All the while, the aim was simply to elaborate Ryon's practices suitable for presentation in a manual. The intentions and accomplishments were at variance.

Humorously so, but tragically true, the USPHS committee-styled researches and the "Joint Committee" authoring the *Manual of Septic-Tank Practice* conformed to Winneberger's Law I [68], "The intelligence of a group as reflected in its activities and decision-making processes decreases as the number of members of the group increases; further, the intelligence of a group decreases as the intelligence of the individual member increases." So much for a collection of experts.

MOST CONSERVATIVE CRITERION

As was said, a straight line with a very slight slope would constitute an envelope curve fitting data presented in Figure 22. Such a fit would conform with research findings.

In a study in 1956 [70], five soils were placed in large lysimeters. Initial, fresh water permeabilities varied enormously. Rates were 29, 10, 6.4, 5.2, and 1.75 ft/day in the five respective soils. After 235 days of sewage spreading, respective permeabilities were 0.1, 0.6, 0.1, 0.1, and 0.1 ft/day. The second, high value occurred because of previous resting of that soil. Essentially all soils were clogged to yield the same permeability after long-term sewage inundation.

The foregoing research findings have a hypothetical explanation. Sewage clogging is an intensive, microbially oriented, long-term reduction in soil permeability. The clogged zone is only a fraction of an inch, or so, thick, at the inundated soil surface [30,61].

Whereas soil characteristics control initial permeability to wastewaters, after long-term inundation the characteristics of the clogging substances themselves control permeabilities. Inasmuch as many soils have about the same porosity, those pore spaces containing clogging substances would offer about the same cross-sectional area in different soils through which the wastewaters must travel.

A second phenomenon affects clogging. Some particle size—about 1 mm—of a porous medium will transmit wastewater solids rather than remove them at the surface [71]. Whereas surface removal of debris results in considerable clogging, penetration (termed in-depth clogging) [17,18], results in reduced clogging. Thus, soils with either strong structure or coarse texture could have a different response to wastewater spreading. Nevertheless, only extremely porous soils would be looked to for such a different repsonse.

Application

In 1965, a proposed subdivision in Alameda County, California, had political problems which required unusual offerings on the part of subdivider and consultant [42]. It was proposed to advise the most conservative disposal fields as seemed practical and to provide a public entity for their maintenance.

Assuming long-term inundation commonplace in disposal fields and that the majority of soils will clog to about the same degree, the Most Conservative Criterion was devised. Review of both Ryon's and the USPHS data disclosed that 0.25 gal/ft^2/day was a loading rate above which all troubled systems were found. Adhering to research findings leading to belief that sidewalls of disposal fields were more effective infiltration surfaces, the value of 0.25 gal/ft^2/day which had been descriptive of bottom areas needed to be applied to sidewall areas. The USPHS did not describe the disposal fields studied, but the likelihood was that gravels were about 6 in. under distribution pipes and 2 in. over them. Each being a troubled field, perhaps all soil surfaces were inundated and functioning as best as possible. The trench could have been 1 or perhaps 2 ft wide. At that time, 0.125 gal/ft^2/day was suggested as a sidewall-only loading rate, but that likely was overly conservative.

In that one subdivision situation, ahead of its time, proposed were:

- concepts of a design criterion based on a single, low loading rate
- dosing on a several-days-per-dose basis
- alternation of two fields on a yearly basis
- a public entity to design, to supervise construction, and to maintain septic-tank systems

That was too much at one time for local authorities who were not yet ready for abandonment of existing practices, however unhappy they were with them.

Since that time, some of those early efforts to carry research findings into the practical world have become reality. Dosing on a several-days-per-dose basis has been found impractical, but alternation of fields has become practice in many areas [45,72,73]. So also has the concept of total management become applied in degrees [46,74,75]. The Most Conservative Criterion did not, as said, find acceptance where and when proposed, but authorities in other California counties saw its practicality.

McGauhey believed the value chosen for the Most Conservative Criterion could be estimated at 0.03 ft/day (0.22 gal/ft^2/day). He assumed a disposal field meeting that requirement being constructed in two sectors, which are used alternately, yearly. McGauhey published his numerical concept of the Most Conservative Criterion first in a Final Report of researches at the University of California [17], which later was published nationwide by the U.S. Department of Housing and Urban Development [18].

In winter of 1967–68, Messrs. L. Raynor Talley, Ed Munson, and William Anderman, directors respectively of environmental health departments of Santa Cruz, Monterey, and San Benito Counties, California, met. They decided to accept the concept of a standardized disposal field as opposed to the classical sizing by percolation test [76].

In Santa Cruz County, in April 1968, a standard septic-tank effluent loading rate of 0.3 gal/ft^2/day was assumed. So also was assumed 75 gpcpd and 4 persons/household of 2- or 3-bedroom houses. Thus, the sanitarians viewed subject soils. If they were acceptable at all, the standard disposal field for 2- and 3-bedroom houses was 1000 ft^2. Only sidewall surfaces were credited in satisfaction of requirements. The percolation test only was used to settle questions of acceptability of soils for septic-tank practices.

Later, in 1972, the effluent application rate was changed to 0.25 gal/ft^2/day in conformance with the first proposed value [42]. At the present time, Santa Cruz County authorities require 600 ft^2 for a 1-bedroom house, and 1200 ft^2 of sidewall area for a 2- to 4-bedroom house. The principle of alternation of fields is followed in that the 1200-ft^2 requirement is divided into two sectors separated by a diversion valve, each of which meets 75% of the overall requirement. Altogether, 1800 ft^2 of sidewall areas (disposal trenches or pits) serving a typical household discharging 255 gpd [3] would amount to 0.28 gal/ft^2/day loading onto one of the two alternate disposal fields, or 0.14 gal/ft^2/day as related to both fields.

Summary

Using history as a guide, the Most Conservative Criterion could be taken to be 0.25 gal/ft^2/day. It is to be applied only to sidewalls of disposal fields. Two such fields are to be provided with a diversion valve such that one field can serve while the alternate rests.

SOIL JUDGMENTS FOR LOADING RATES

In 1952, Federick [66] noted, "It must be recognized that the majority of subsurface tile fields are installed by builders who do not perform percolation tests." That situation still holds today. Consequently, efforts have been made to offer visual descriptions of soils as they might relate to percolation rates and minimum design sizes.

Ryon's Soil Judgments

When Ryon performed his studies, he judged and described soils in his own terms [22]. A table was presented in his early work in which soil descriptions were related to the length of tile (square feet of bottom area of trenches) needed per person. Ryon estimated water usage at 80 gpcpd. Later, Ryon's colleague, Joseph Federick, published Ryon's unpublished notes, giving more details [66].

Taking Ryon's work and Federick's presentations of Ryon's unpublished notes together, Table XII was assembled. There, Ryon's loading rates are presented in terms of bottom areas, which Ryon did. Loading rates are also presented in terms of sidewalls plus bottom areas, both being treated as equally functional surfaces.

It is interesting to note that Ryon's table provided for soils with percolation rates much slower than 75 min/in., being far off-scale of Figure 21.

UPC Soil Judgments Compared to Ryon's Practices

Authors of the *Uniform Plumbing Code* (*UPC*) presented no percolation test procedure. Soils descriptions were preferred.

Neither Ryon's nor the UPC's descriptions conformed to any formal schemes of soil classification. There is essentially no way a technician could view a soil and recognize it in terms of Ryon or the *UPC*. To provide a frame of reference, Table XIII was assembled. Percolation rates corresponding to loading rates (Bottom Areas) were taken from Federick's formula as presented in the *Manual of Septic-Tank Practice*.

Table XII. Ryon's Soil Descriptions Related to His Practices[a]

Soil Description	Design Factor (ft²/person)	Loading Rate (gal/ft²/day) Bottom Area Range	Bottom Area Mean	Functional Area Range	Functional Area Mean	Percolation Rate (min/in.)
Clean coarse sand	15– 25	5.33–3.20	4.27	2.28–1.37	1.83	1/6– 2
Fine sand, or light loam	25– 36	3.20–2.22	2.71	1.37–0.95	1.16	2– 5
Sand and clay or loam	36– 48	2.22–1.67	1.95	0.95–0.72	0.84	5– 10
Clay with some sand or gravel	90– 180	0.89–0.44	0.67	0.38–0.19	0.29	25– 70 +
Heavy clay	350–1000	0.23–0.08	0.16	0.10–0.03	0.07	120–600

[a]Design Factor assumes 80 gpcpd. Loading rates were calibrated to a trench 1 ft wide and with 8 in. of useful sidewall height. Functional Area relates to bottom plus sidewalls.

Comparing Tables XII and XIII, one sees quite similar soil descriptions but that is the end of similarities. Note, for example, that whereas Ryon described "heavy clay" as having a percolation rate range of 120–600 min/in., the *UPC* described "clay with small amount of sand or gravel," which had a corresponding USPHS percolation rate of 20 min/in. Not only does the *UPC* seem less conservative than Ryon, but many local authorities follow the USPHS percolation test, which is less conservative than Ryon's test in the less permeable soils.

U.S. EPA Soil Judgments

In an effort to provide soils descriptions in formal terms as related to loading rates of disposal trenches, the U.S. Environmental Protection Agency (EPA) provided a table in which soils were described in terms a soil scientists would use [16]. That table (Table 7-2, Ref. 16) less its loading rates is presented herein as Table XIV.

Despite the improvement in terminology in Table XIV, however, the range of percolation rates corresponding to soil textural types is much off-scale. Note Table V (Chapter 4, page 34), for one example. Study No. 1 was performed in soils classified by laboratory analyses as sandy loams. Yet whereas the U.S. EPA assigned percolation rates to "sandy loam, loam" at 16–30 min/in., Study No. 1, replicated 36 times, had a range of 2.6–78 in/hr. (~0.8–23 min/in.) and a geometric mean rate of 11.7 in./hr. (5.13 min/in.).

Without going through the remainder of the studies in Table V and through years of accumulations of data, the writer summarily asserts that the U.S. EPA Table 7-2, Ref. 16 (Table XIV herein) fails by a huge margin to conform with his experience of personally performing thousands of percolation tests and mechanical analyses of soil texture of soils from one coast to the other.

Table XIII. Uniform Plumbing Code Soil Descriptions Related to Practices[a]

Soil		Design Factor	Loading Rate ($gal/ft^2/day$)				Percolation Rate	
			Bottom Area	Functional Area Basis				
Type	Description	(ft^2/person)	Basis	1½ ft Wide	2 ft Wide	3 ft Wide	(in./hr)	(min/in.)
1	Coarse sand or gravel	20	5	2.14	2.50	3.00	60, >60	(1, <1)
2	Fine sand	25	4	1.71	2.00	2.40	38	(1.6)
3	Sandy loam or sandy clay	40	2.5	1.07	1.25	1.50	15	(4.0)
4	Clay with considerable sand or gravel	60	1.66	0.711	0.830	0.996	6.6	(9.1)
5	Clay with small amount of sand or gravel	90	1.11	0.475	0.555	0.666	3.	(20.)

[a]Design Factor assumes 100 gpcpd. Percolation rates assume the USPHS *Manual of Septic-Tank Practice*; $Q = 5/\sqrt{t}$, where $Q = gal/ft^2/day$, and $t =$ min/in. Functional Area assumes a UPC trench with 1 ft of useful sidewall heights.

Table XIV. U.S. EPA Relationship of Soil Descriptions and Percolation Test Rates (After the U.S. EPA [16]).

Soil Texture	Percolation Rate (min/in.)
Gravel, coarse sand	<1
Coarse to medium sand	1–5
Fine sand, loamy sand	6–15
Sandy loam, loam	16–30
Loam, porous silt loam	31–60
Silty clay loam, clay loam	61–120

Summary

No table relating percolation tests can successfully be well related to visual soil descriptions.

Muskat [77] studied movement of water through soils. His remarks bear well on efforts to relate soil descriptions to percolation rates:

> Although it is not inconceivable that there may ultimately be found a relation between the permeability of a porous medium and the porosity, grain sizes, their distribution, and a shape factor for the grains, it is at present certainly far simpler and more accurate to make a direct permeability measurement by means of a flow experiment than to carry through a laborious sieve analysis and a porosity determination and then make a permeability calculation by means of a relation that has at best only a limited range of approximate validity.

If a percolation test program cannot be undertaken and history of existing systems in the locale cannot serve as a guide, it is recommended that conservative design criteria be applied, assuming soils appear suitable at all.

TIME LOADING A DISPOSAL FIELD

The long-term affect of inundating soil with freshwater, Phase III, was discussed (Figure 18). Eventual clogging was attributed to microbial activities. The presence in wastewaters of organic matter—foodstuffs for microbes—magnifies clogging (Figure 19).

On the other hand, microbes have been found beneficial to soil permeability through production of substances leading to soil aggregation [78,79]. Adding organic foodstuffs helped permeability, but a period of water spreading for incubation followed by a drying out period was needed [80]. Thus, it seemed that disposal fields receiving quantities of organic debris might under most circumstances clog readily and under other circumstances recover permeability. Certainly a period of incubation is ordinarily provided, but the chance to dry out seems hazardously dependent on the capricious inclinations of households towards extended periods of absence and cessation of wastewater flows.

Later, it was found [48] that partial recovery of the initial infiltrative capacity of a sewage-clogged soil did not require complete drying. A resting period during which soils were drained and aerated, although still quite wet, was adequate. It was further found that the loss and gain of permeability was a matter of balance of anaerobic and aerobic processes. That corresponded with information gained from studies of freshwater spreading basins. Thus, the finding that time loading and resting, rather than volumetric loadings as such, have a stronger affect on soil clogging processes, had a rational explanation.

Dosing

Knowing that time was an important factor in loading a disposal field with septic-tank effluents led investigators to look to dosing as a possible method for translating research information into field practices [41]. It was hoped that draining and resting soils between inundating doses would be beneficial. Experiments with resting effluent-inundated soils, however, terminated those hopes. It was found with a very permeable soil—Oakley sand—which would readily drain and begin resting, that beneficial effects of dosing would require a rest of several days. Clayish soils, because of slow drainage, would likely require even longer resting periods between doses. Hence it seemed that any genuinely effective scheme of dosing would require a tank capable of holding household wastewaters for several days. One investigator, McGauhey, noted, ". . . time for full restoration of lower horizons is prohibitively long for application to each daily cycle of loading and resting" [17,18]. The other investigator, the writer, found in private practice that dosing was impractical. A dosing arrangement both complicated septic-tank systems and constituted costs, which monies could far more usefully be placed in additional disposal field capacities.

Had dosing been found a worthwhile theoretical objective, which it was not, the investigators still foresaw practical difficulties. Not all soils were as permeable as one would wish, and draining completely to begin rest in all parts of a disposal field before the next dose came seemed universally improbable. Also, bottom surfaces, horizontal as opposed to vertical, recovered haphazardly and slowly from periods of effluent inundation. It seemed obvious that some low spot in a disposal field would be horizontal and last to drain. The time would come when the next dose would preclude resting of that spot. As time passed, a higher increment of soil surrounding that low spot would succumb to a permanent pond. Progressively, the pond would rise and the purpose of dosing would be totally defeated.

As said, dosing arrangements, to be effective, would require several days between doses. As Figure 19 shows, within a 24-hr period of time, less than about 3% of the initial percolative capacity of the soil remained. Dosing even on a daily schedule has more losses of permeability to undo than is reasonable to expect. As will later be discussed, clogging could be regarded as acceptable.

By the bye, some authorities are requiring dosing arrangements for disposal fields in clayish soils as well as everywhere else within their spheres of influence. Those dosing arrangements are generally scheduled only for a portion of a day. Disposal fields in many soils cannot drain to empty in less than a few days. It is manifestly pointless to dose to a ponded disposal field.

Alternation of Fields

It had been long known by contractors that a good method of repair of a disposal field was to construct a new field and provide arrangements for diverting septic-tank flows to the new field. The homeowner was told to change the flow to the old field later on. Thus, use of two alternate fields had been empiric practice. Recommending alternation of fields seemed theoretically sound and practical [17,18,41]. Experience has borne out convictions [45,72,76].

The concept of alternation of fields (dual systems, etc.) has been translated to field practices through a rather slow evolution. Nothing changes rapidly in the sewage management world. It must change organically. Alternation of fields on a yearly rotational basis was accepted in recommendations for practices within subdivisions. Soon thereafter, and concurrent with much proselytizing, some of California's directors of environmental health applied the concept as standard practice. The concept has been applied in other states too. (See Figures 23 and 24.)

Regardless of technical merit of the principle of alternation of fields, there are four aspects that have rendered it practical. Firstly, a local authority wishing to increase disposal field capacity in his area often finds

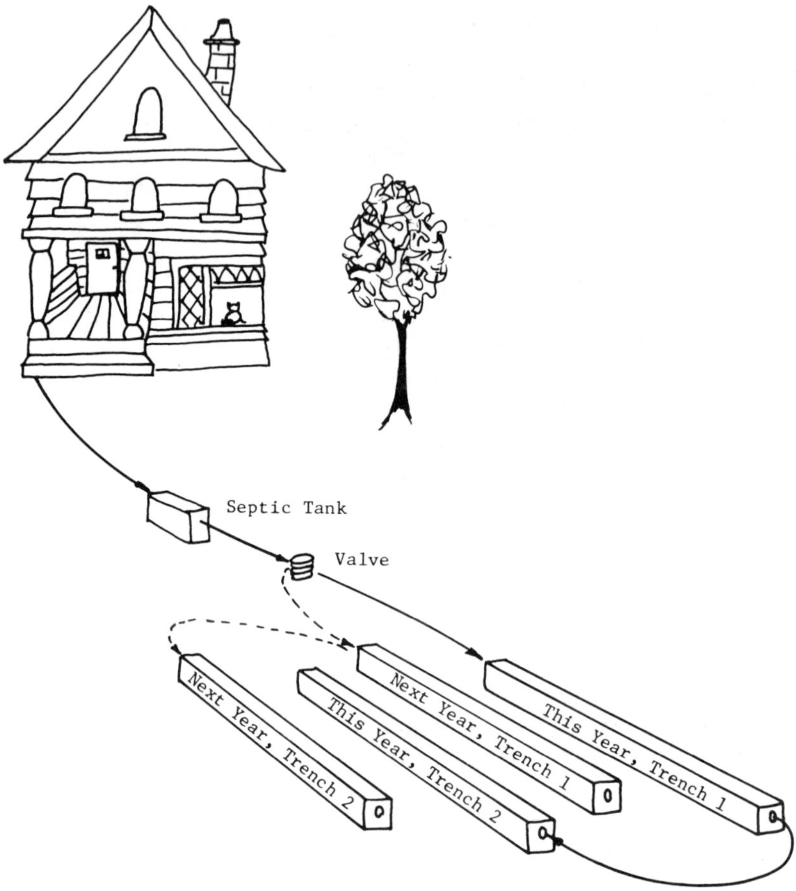

Figure 23. Alternation of interlaced fields.

the concept of resting overworked fields politically acceptable. Most people have had experience with resting things to increase their lifespans. So, local authorities have preferred not recommending splitting an existing requirement into "half-sized" fields. Rather, they recommended overall increases. Some have essentially doubled the local requirements overnight by recommending two complete, alternate fields.

Secondly, when troubles come to ordinary disposal fields, odors and nuisances are manifest. They persist until a repair can be constructed. The presence of an alternate disposal field, even where a homeowner failed to change the valve yearly, offers an instant repair. There is no need for sewage to be on the ground for extended periods of time.

In California, a local sanitarian pointed out a recurring failure. Sewage

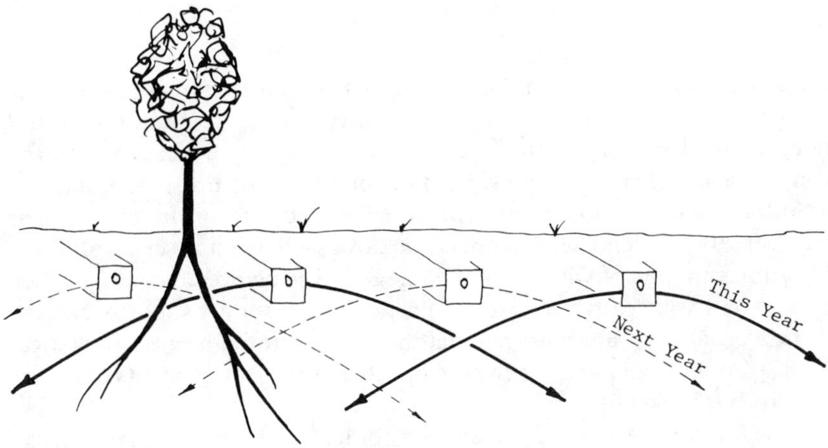

Figure 24. Alternate use of interlaced fields always subirrigates the tree, although from different sides.

was again flowing down the street. It was the season of rains. The prior repair had been made a year ago in wet soils. Advised to ignore the situation until dry weather came and soils were drier, a subsequent repair was successful. The problem, which could have been avoided by presence of an alternate field at the onset, was destruction of soil permeability by digging soils when wet.

Thirdly, hardware to instrument alternation of fields has reached the market [81,82]. Although the concept of alternation of fields apparently had been common practice in Ryon's day—the mid-1920s—methods of diverting effluent flows had been a matter of improvisation. Sometimes guillotine gates in a distribution box were used. In recent times, plastic distribution pipes to disposal fields may have a protruding end into a distribution box. Effluents can be excluded from a pipe by either stopping it with a rubber ball or by slipping a plastic ell fitting onto the pipe with the tail end up. Sanitarians and homeowners are more likely to prefer valves that do not expose the operator to sewages.

Fourthly, repairing a disposal field at a later date can ruin garden areas, landscaping, sidewalks, and patios. Sanitarians and homeowners frequently see value in having a future need met at the onset. The sanitarian will see an advantage to having an installed, second disposal field. Its presence would prevent a homeowner from using the reserve area for incompatible improvements later. From the practical point of view, "reserve area" is an initial concept which has no enforceable reality later on.

A Technical Thought

The question is validly posed, "Why have alternate fields of full size each, when a field twice as big as required would perhaps last about indefinitely?" Aside from the foregoing discussed side benefits of alternation of fields, there is a consideration of soil structure rehabilitation.

Soil structure, upon which most soils depend for much of their permeability, is delicate. Compaction can result from a series of slight vibrations, no one of which could damage soil structure much [83]. When soils have a wide range of sizes of particulates, as most soils do, smaller particles may migrate with percolating waters through large interstices [84,85]. And, small particles freed from structural arrangements clog very effectively [39,40,86].

Soil structure can be weakened without its immediate destruction [38], but it is highly likely that soil structure would be damaged during many years of about constant effluent inundation in a disposal field. It would be expected that periodic drying, such as a year's rest of an alternate field could provide, would be beneficial.

LONG-TERM ACCEPTANCE RATE

The question arises, does the clogging (depicted in Figure 19) progress indefinitely until zero infiltration occurs? A delineating experiment would last at least half a century. No one has done that. It is the writer's belief that zero infiltration is not reached, but that clogging continues slowly until extremely low infiltration rates are had. So far as design of disposal fields is concerned, a designer wishing a reasonably permanent facility needs to know what a practical long-term infiltrative capacity of a soils system is likely to be.

Aside from common experiences, the USPHS measured survival curves in many soil types and localities in our country. There is little doubt that the ordinary disposal field has a lifespan. And, indeed, most people involved with septic-tank systems believe that a lifespan is unavoidably intrinsic to those systems. So much so are authorities convinced that the lifespans are inevitable, unpredictable, and too short, that they commonly present recommendations for septic-tank practices with a preface amounting to, "The best method of disposing of household sewages is to connect to a community sewer, but if you must have a septic-tank system, here is our guidance for those temporary devices."

It should be obvious that disposal fields can last forever. It is only a matter of proper size. Does the reader doubt that a disposal trench reaching across Kansas on a contour would not serve a single-family residence indefinitely?

Professor Rein Laak, his colleagues, and students have constituted a leading center for research on septic-tank systems. Experimental arrangements are with genuine sewages, septic-tank effluents, and the like. (Studies with simulated sewages are, the writer asserts, about as invalid as studies of simulated marriages.) Out of those studies, the concept of the Long-Term Acceptance Rate (LTAR) has evolved [6]. A basic premise is the belief that a sewage-clogged soil surface offers better treatment of infiltrating wastewaters, and systems can be designed on the low permeability a sewage-clogged soil offers.

Laak [6] and Healy and Laak [87] have proposed use of a falling-head permeameter for measurement of soil permeability. Because great latitude is offered by application, simple permeameters offer sufficient accuracy. Darcy's coefficient of permeability has been related to the LTAR, the loading rate of sidewalls plus bottoms of disposal fields:

$$\text{LTAR} = 5k - (1.2 \div \log k)$$

where k is Darcy's coefficient of permeability in ft/day; and the LTAR has units of gal/ft^2/day.

Using the correlation suggested for relating coefficients of permeability to USPHS percolation rates (Chapter 3, page 30), and transforming units for dimensional consistency, estimated LTAR values have been calculated and are presented in Table VIII (Chapter 7, page 79).

Laak has pointed out that the LTAR has best application between 0.2 to 0.8 gal/ft^2/day [88]. "Over 0.8 gal./sq. ft./day the field data showed so-called 'unstable' values—too much variation. . . . Very few data points were available for LTAR values less than 0.2 gal./sq.ft./day. . . . (The lowest LTAR values for septage lagoon bottoms are about 0.1 gal/sq.ft./day.)"

As shown in Table VIII, LTAR values appear more conservative than Ryon's loading rates translated into functional areas, up to about 30 min/in. percolation rates ($k = 1.0 \times 10^{-4}$ ft/min). With slower percolation rates, Ryon's loadings appear lesser. Again it should be pointed out that Ryon deviated from his curve in the less permeable soils. Figure 22 gives the curve Ryon actually followed in his recommendations. One sees the departure from the "line used" to Ryon's extrapolation of the "line including all points."

The Most Conservative Criterion of 0.25 gal/ft^2/day is not in much variance with lower LTAR values. Whereas the Most Conservative Criterion ignores soil permeability beyond acceptability at all, the LTAR credits more permeable soils. Recall an earlier comment that an envelope curve for Figure 22 would be a straight line running from percolation and loading rate coordinates from 4.2 min/in. - 0.49 gal/ft^2/day to 111 min/in.-0.254 gal/ft^2/day. That line agrees surprisingly well with corresponding LTAR points given in Table VIII. Also note that LTAR values have their valid range (0.2–0.8 gal/ft^2/day) in good agreement with the percolation rate range one generally encounters in field practices.

Comments

The writer has seen too many septic-tank systems constructed, to believe that intended loading rates find their way to accurate reality. Not only are wastewater flows highly variable, but the geometries of disposal fields vary much from intentions.

Experience is that percolation rates are very generally known only with very crude accuracy. Even with careful techniques, it is generally impractical to replicate the percolation test as many times as needed for much accuracy. And, local authorities are quite reluctant to change practices from tradition. Anyone who has long dealt with the public would well understand the discomfort an authority would suffer in proposing change to a proper percolation test, while being asked what was wrong with his old advised test before.

Therefore, it seems best to advise a reader to select from the information offered herein whatever system he believes to be a good one and which he knows to be acceptable in his situation.

CHAPTER 8

DISTRIBUTION TO DISPOSAL FIELDS

Perforated distribution pipes could be smaller and less expensive, could be laid almost anywhere within gravel fillings of disposal fields, and technically need not conform to some specified grade.

Each year, valuable man-hours are wasted laying gravity-flow, perforated, septic-tank distribution pipes either to level or to some specified grade. Also, some authorities, believing that the pipe must be 4 in. in diameter and needs to be laid straight, have demanded rigid pipe where a flexible perforated pipe and smaller-than-usual diameter pipe could have served as well.

Where a disposal field is to be served by gravity flow, it is asserted:

1. Within the gravel filling of a disposal trench, bed, or pit, one perforated pipe running between the most extreme ends, or corners, could serve. In a trench, the pipe could run from one end to the far end, and the grade and straightness would be inconsequential. In a wide bed, a single pipe could run diagonally from one corner to the far corner. More pipes could be used, but they would add little or no benefit. In a pit, where pipes are not always required, a pipe should be used, and it should run from the top of the pit to the bottom.
2. It would be best to provide about 4 in. or so of gravel filling between the perforated pipe and any part of the soil, including the soil backfill overhead. The intent is to discourage plant roots from entering and clogging the pipe.
3. Rather than waste any of the gravel-filled excavation, see to it that the uppermost level of the gravel filling is at or below the level of the invert of the septic-tank outlet. There is no reason why gravel-filled areas above the bottom of the pipe cannot eventually be useful for infiltration of effluents.
4. Perforated pipes could be either flexible or rigid, and for practical reasons could have diameters not much smaller than 2 in., and pipes should have perforations about the size they now generally are when gravity flow is anticipated.

The foregoing statements do not conform with practices in many areas, and many readers would appreciate a presentation of some of the reasons for the foregoing assertions.

NEED FOR PERFORATED DISTRIBUTION PIPES

Gravels used within disposal fields are customarily quite coarse. Septic-tank effluents devoid of coarse solids flow through gravels freely. At first thought, one might believe that perforated distribution pipes would not be needed. Indeed, some authorities only require a solid pipe to a disposal pit, and a perforated pipe running from top to bottom of the pit is omitted.

Distribution pipes are needed for hydraulic continuity throughout the gravel-filled element of a disposal field and for defeating gravel clogging from sludge carryover from ultimately neglected and poorly designed septic tanks.

Continuity For Effluent Flow

Authorities demand clean gravels for disposal fields and their drawings always depict gravels neatly placed in almost perfectly dug excavations. Still, there is the practical world of contractors and construction methods with which one must reckon.

When leaving the University of California to enter private practice, the writer believed a consultant on septic-tank systems should understand the contractor's viewpoint. Accordingly, he became a licensed contractor and contracted some work. So from that early contracting experience and from watching many other contractors construct septic-tank systems over the years, it is asserted that it is not realistic to expect a contractor to leave much valuable gravel at the bottom of his pile when placing gravel into an excavation with his front-end loader. He usually scoops up some soil with the last of the gravel, and sometimes the soil quantities included are considerable. Consequently, a chance exists that soil included with gravel could constitute a barrier separating one part of a disposal field excavation from the other.

Soils vary much in their cohesive strengths. Some rigidly maintain excavation integrities, while others do not. Silty soils and sands especially tend to migrate into gravel-filled excavations.

For the foregoing reasons, one useful purpose of having a perforated distribution pipe is for ensuring a continuous pathway for effluent flow from one end of an element of a disposal field to the opposite end.

Openings to Defeat Pipe Clogging from Sludge Carryover

Most septic tanks in common use are poorly designed. Very commonly, homeowners neglect needed periodic cleanings. Consequently, authorities demand ever larger septic tanks, perhaps hoping for a tank large enough to contain a lifetime's accumulation of scum and sludge. Still, as decades of neglect pass, a neglected septic tank will ultimately discharge scum and sludge to the distribution pipes of the disposal field.

Except for the discussed need for hydraulic continuity from one end to the other end of the disposal trench (bed or pit also), a pipe carrying solid-free septic-tank effluents to the trench would serve quite well if simply stubbed into the gravel. But when some scum and sludge carry over from the neglected septic tank, the few gravels facing the blunt end of the pipe would soon clog. After all, the open blunt end of the 4-in.-diameter pipe faces only about 12.5 in.^{2} of gravel surface.

Thus, it is a distinct benefit to have a perforated pipe from one end of the trench to the opposite end. The multiplicity of perforations along most distribution pipes offers far greater total area opening to gravels than the diameter of the pipe itself has. Consequently, the perforated pipe tends to delay ultimate gravel clogging from scum and sludge carryover.

HISTORIC MANAGEMENT OF DISTRIBUTION PIPES IN DISPOSAL TRENCHES

Frank and Rhynus, 1919

Probably the first USPHS investigations of septic-tank systems [89] led the investigators to recommend use of two disposal fields separated by a control chamber in which a flat board "gate valve" permitted one field to be used while the alternate field rested. The principle of alternation of fields [72] was early appreciated.

Frank and Rhynus advised distribution pipes consisting of 3-in.-diameter agricultural drain tiles, laid directly on the bottom of an 8-in.-wide trench, embedded in gravel to the top of the tiles, covered with a board laid directly on top of the tiles, and backfilled with about 6 in. of soil to grade.

As for grade of distribution pipes, 2 in./100 ft (0.17%) was an advised downslope gradient, and drawings depicted use of a dosing tank and automatic siphon.

Henry Ryon, 1926

Ryon, working for the New York State Department of Health, advised drain tiles to be sized from 3 to 6 in. in diameter and to be laid on a slope of 4 in./100 ft [22]. Trenches customarily were 12 in. wide and with 2 in.

of gravel under the pipe and with 2 in. over it. Possibly Ryon conceived of disposal trenches as being soil drains working backwards. Ryon commented, "Slopes too steep will overload the portion of the field at the extremities of the tile lines and cause the sewage to rise to the surface. . . . Slopes too flat will overload the upper ends of the tile lines."

New York State regulations in 1940 [90] advised 4-in.-diameter pipes to be laid on slopes not greater than 0.5% (6 in./100 ft).

Metcalf and Eddy, 1935

Leonard Metcalf and Harrison Eddy have long been respected names in the field of sanitary engineering. In their early text, Metcalf and Eddy [91] noted, "The tile used in subsurface irrigation systems may be either unglazed drain tile or glazed tile, 2 to 6 in. in diameter." The currently popular belief that distribution pipes must be 4 in. in diameter fails to conform with history.

Metcalf and Eddy further noted. "In some cases the entire pipe is surrounded with graded gravel, extending two inches below and an equal distance above the pipe and for the full width of trench. The slope of the lateral pipes depends somewhat on the porosity of the soil. In porous material, it is generally greater than in more impervious soil, so that some sewage may run all the way to the lower end of the system instead of all of it flowing out of the first few openings. In general, the slope is 2 to 5 inches per 100 feet. . ." [0.17–0.42%].

Although residential and institutional systems were the subject of Metcalf and Eddy's text, the latter and larger systems were the focus of attention. Dosing tanks and siphons were advised, and indeed there are hydraulic problems of distribution peculiar to large systems.

U.S. Public Health Service, 1957

By the middle of this century, tradition had become such that the USPHS *Manual of Septic-Tank Practice* [92] advised that distribution pipes consist of, ". . . 12 inch lengths of 4-inch agricultural drain tile, 2 to 3 foot lengths of vitrified clay sewer pipe, or perforated, nonmetallic pipe." What the national authority advised, of course, became law in many places. So, the misconception that distribution pipes had to be 4 in. in diameter became widespread.

In municipal sewage treatment plants, one popular treatment has consisted of trickling sewages over rocks. That aerobic process became confused in many people's minds with the flow of septic-tank effluents (anaerobic) over gravel in disposal fields. Perhaps for that reason—for none other seems apparent—the USPHS pictured distribution pipes no longer closer to the trench bottoms as they had been earlier. (It is always tempting to guess the reason for a governmental mandate; nevertheless, there doesn't have to be a reason.) Rather, the USPHS suggested 6 in. minimum of gravel under the distribution pipe and only 2 in. above. At the same time, investigators of the USPHS found that when plant roots entered distribution pipes, the pipes tended to be either in contact with soil or too near it (gravel shorting). Recommending only 2 in. of gravel over the pipe has likely resulted in many root-clogged distribution pipes. Consequently, about 4 in. of separation of the pipe from any soil surface would be a better rule.

That first printing of the *Manual* advised distribution pipes to be laid at a grade of 2 to 4 in./100 ft and never to exceed 6 in./100 ft.

U.S. Public Health Service, 1967

After many investigations of disposal fields, the USPHS revised the *Manual of Septic-Tank Practice* to conform with changed thoughts [15]. As for slopes of distribution pipes, prior advice was revised, "The bottoms of the trenches and distribution lines should be level. . . . The bottom of the bed and distribution tile or perforated pipe should be level."

The Uniform Plumbing Code

The *UPC* [47], product of a committee of the International Association of Plumbing and Mechanical Officials, is (to date) in variance with research findings of the USPHS. The *UPC* advises "approved pipe." Minimum and maximum slopes of distribution pipes are set at 3 to 6 in./100 ft. Distribution boxes, well known sources of troubles in sloping terrain, are also mandated to be used. Following the *UPC* doubtless results in unnecessary, premature failures.

A Practice

During a brief while between the use of the tiles and perforated plastic pipes, a perforated bituminous paper or fiber pipe was used in places. Although those bituminous pipes could be laid with fittings, more than one contractor, in making a turn for example, would chop two pipe ends with a hatchet to a very rough angle. Those were butted. Ends of pipes were customarily left open. Ends of perforated plastic pipes are also often left uncapped. Even distribution out of perforations could not have been even an imaginary happening.

Conclusions

Although septic-tank effluents smell like sewages, experienced people know that they are far more like clear water to manage in pipes. Whereas ordinary sewages will clog sewer pipes without proper slopes, septic-tank effluents free of large solid debris do not. Still, there has been a tendency of engineers to think in terms of sewages complete with solids. So, perhaps from habit, they began prescribing a variety of slopes for distribution pipes in septic-tank systems. But not all of the variety could be right. And because disposal fields through the years have by and large functioned properly with a variety of pipe slope requirements, it seems reasonable to assume that slope, as such, mattered little, if at all.

REGULATIONS AND ENERGY WASTES

The United States has a history of popular movements which come and go without many constructive accomplishments—prohibition, antianarchists, anticommunists, save the environment, save water, and so on.

The latest—energy conservation—would best be dubbed the "energy conversation movement." Little is done to save energy, besides pricing it beyond the reach of the less affluent. Rather, the efforts seem to be in finding more sources of energy to exploit. Recognition of the oversized distribution pipes and revision of codes could save energy.

In 1977, it was reported that about 20 million lineal feet of plastic pipe were yearly installed in disposal fields in the United States [93]. Those 4-in.-diameter pipes were fabricated of about 9.5 million lb of plastics. Smaller, 2-in. pipes have been asserted to be adequately large for disposal fields [32], and their replacing the larger pipes would require perhaps 2.6 million lb of plastic, a savings of 6.9 million lb.

Those oil-derived plastics could be viewed for what they are—energy. Each pound of polyethylene plastic requires 12,850 Btu to produce. Added to the heat of combustion, the energy that plastic represents is 34,490 Btu/lb.

Using polyethylene as a baseline, the savings to be realized by use of smaller diameter distribution pipes would be about 210 billion Btu yearly.

SLOPE OF DISTRIBUTION PIPES

Foregoing remarks of Metcalf and Eddy clearly point to belief that some value was to be had by spreading effluents evenly throughout disposal fields. That, in turn, implies belief that each perforation of a pipe (or butted tile joint) will discharge its proportionate share of an effluent volume.

Perforated Pipes are Customarily Huge

An ordinary household might discharge 255 gal of wastewaters daily [3]. Calculations show that daily wastewater flow would fill a 390-ft-long pipe of 4-in. diameter only once. Intuition alone tells us that pipes of 4-in. diameter are oversized for the job to be done, and the chances of effluent trickling evenly out of the multiplicity of perforations are infinitesimal indeed. If equal distribution of effluents through all perforations is not achievable, laying a pipe to a specified grade seems particularly pointless.

Trying to Level Perforated Pipe Hydraulically

A few years ago, while constructing a disposal bed to be dosed with septic-tank effluents by a pump, a K & E contractor's level was used to level rigid perforated pipes as accurately as possible. Meanwhile, the septic tank and dosing tank were filled with fresh water.

The pipes were charged with water from the pump. Water dribbled out of a couple of perforations at one pipe end, a perforation towards the middle of another pipe, and so on. Then, using the water as an indicator of hydraulic level, the pipes were further adjusted. After an hour of efforts, it became quite obvious that no amount of adjustments would achieve anything near even distribution through all perforations.

At the University of Connecticut, Milan Knor measured volumes of discharges from perforations along a 75-ft-long pipe [94]. Even distribution did not occur.

Intuitively, realizing the futility of trying to have all perforations at an exact elevation and to have distribution pipes absolutely straight on some specified grade, only one contractor known to this writer wholeheartedly tries. Many do not. Despite inspection programs, there is no practical way for authorities to make a contractor really try to achieve something he believes to be senseless.

Progressive or Creeping Failure

As stated, the USPHS recommended distribution pipes to be about level [15]. That recommendation resulted from understandings gained from extensive in-field investigations of residential septic-tank systems.

James Coulter, an investigator of the USPHS, noted, "Our studies lead us to the conclusion that soil clogging is a progressive phenomenon. . . . A given increment of the total absorptive capacity will be exhausted each year regardless of the method of distribution" [20].

At the University of California, studies of clogging of soils with septic-tank effluents led McGauhey to describe "creeping failure" as an expected occurrence [17,18].

In a disposal trench receiving a septic-tank effluent, regardless of just where in a trench the effluent discharges through perforations of a distribution pipe, the incoming effluent seeks its own level rapidly in the gravel filling. At first the lowest spot gets about all of the effluent and that spot clogs and ponds. Later effluent flows add to the pond and fresh soils surfaces become inundated, also only to clog as the pond creeps upwards. Thus, increment by increment, soil surfaces clog upwards. That is progressive, or creeping, failure. The word failure is an unfortunate choice because residual permeabilities of clogged soils can serve for infiltration for years to come.

The concept of progressive, or creeping, failure implies that just where effluent trickles into the gravels of a disposal field would be inconsequential to the outcome. In turn, that implies that slopes of distribution pipes are inconsequential as well.

Steps Forward

In 1973, The State of Arizona recognized need to update septic-tank practices to conform with knowledge [32]. New regulations were provided, "To improve understanding and operation of septic-tank systems." The Arizona State Department of Health Services provided "Guidelines for Installation of Septic Tank Systems," [95] which noted, "Minimum diameter of the distribution pipes shall be two inches." Sales being small in Arizona compared to some of the remainder of the states, industry has yet to provide a properly perforated 2-in.-diameter distribution pipe for sale in Arizona, but the way was opened.

Authorities in Arizona also provided, "Both distribution pipe and trench bottom shall be approximately level," and, "Rock or gravel fill shall extend from the bottom of the trench to four inches above the distribution pipe." In recognition that plant root invasions are discouraged by separation of the pipe from soils, Arizona authorities further noted, "Distribution pipe shall be a minimum of four inches from any soil surface (sidewall or soil cover)."

SERIAL AND PARALLEL DISTRIBUTION

It seems to be a natural inclination to pity the first trench in a serially distributed system. It is loaded to capacity before other trenches are loaded, and it remains loaded during service of other trenches. Nevertheless, serial distribution is superior on sloping terrain, and it is as good as, or superior to, parallel distribution in flatlands. In answer to an inquiry, Coulter of the USPHS noted, "Our calculations show that the clogging in a serial trench would need to be 2 to 50 times as great as that in a distribution box trench (parallel distribution) before a perfect distribution system would begin to have merit." [20].

The concern for the first trench in a serial system has its roots in belief that it is loaded to "failure."

According to Laak [6] and experience with long-lived disposal fields, soil clogging reaches a fluctuating equilibrium between forces tending to clog more and forces tending to unclog. The first trench then is only overloaded hydraulically, thus the need for the remainder of the system.

One may view disposal trenches as individual units, but in function they are united. Assume, for example, a system of four trenches happens to be constructed such that the long-term performance will accommodate a permanent pond requiring half of the total sidewall area. If that system has parallel distribution, all four trenches will be ponded halfway up the sidewalls. The lowermost soils serve longest, while uppermost soils wait their turn for surges or ultimate use, if ever. If that system has serial distribution, the first two trenches will accommodate the pond, while the third and fourth trenches, constituting half of the total sidewall areas, wait their turn for surges or ultimate use, if ever. Aside from considerations of hydraulic head, superior in the serially loaded trenches, each method of distribution results in constant inundation of soils amounting to half of the total sidewall areas.

GRAVEL

In addition to shoring sidewalls and otherwise maintaining integrities of excavations, gravel fillings are a major part of the distribution system. Once out of the perforated pipe, effluents must be transmitted through gravel to reach soil surfaces.

During field investigations, the USPHS found many neglected septic-tank systems. People generally did not clean (pump out) their tanks. It was found that sludge carryover could pass through coarse gravel to reach lower reaches of disposal fields. Avoiding clogging the distribution pipe before clogging the trench seemed to the USPHS investigators to add a few more years before total neglect would terminate the useful lifespan of the disposal field. It appeared that coarse gravel, as opposed to finer, offered a benefit.

At the same time, investigators found silty soils intruding coarse gravel fillings, often up to the distribution pipe, to form a useless slurry of gravel and soil. In such soils, finer gravel, or perhaps coarse sand, would have served better than coarse gravel in maintaining the integrity of soil excavations.

Faced with the decision to advise national practices on gravel size, authors of the USPHS *Manual of Septic-Tank Practice* decided that there were far more negligent people with septic-tank systems than there were cohesionless, silty soils. So, very coarse gravel was advised.

Where there is reason to expect proper maintenance, smaller gravels are better for supporting anything as fine grained as soils are. Additionally, finer gravel (perhaps 0.5 × 0.5-in. or smaller pea gravels) will better slow surge flows. The energy of surge flows would be better spent within gravel than against infiltrative surfaces of soils.

Crushed limestone is sometimes used for gravel fillings of disposal fields. Limestone consists of carbonates of calcium and perhaps magnesium, and minor constituents. Depending on the mixture and hardness, limestone dissolves more or less easily in acids. Septic-tank effluents contain hydrogen sulfide which will be converted to dilute acids capable of attacking some limestones.

Rouman [96] has reported his and sanitarians' experiences viewing limestones in disposal fields. Apparently, limestones dissolve and resolidify into solid masses, ". . . or at least cause failure of the system." As best as Rouman could determine, ". . no local health department in Michigan allows it (limestone) to be used and the Michigan Department of Public Health does not recommend its use."

Gravelless Backfills

Where gravel has been in short supply or quite costly, other materials have been used. In Arizona, for example, occasionally a pit was dug. An old automobile body was thrown into the pit. Then a distribution pipe (or building sewer) was stubbed into a broken window, and the excavation was soil backfilled. That and similar arrangements served in those particular soils situations and human affairs. Most people would be quite surprised to see the collection of junk that finds its way into disposal fields.

For some years, Fred Crates has been developing the use of corrugated plastic pipes in backfilling disposal fields. Installations in areas where gravels were practically unavailable have served quite well for some years. Even though plastic pipes may be a bit more costly, Crates reports that the time and labor required for handling gravels often make use of plastic pipes competitive [97].

There is a need to stop damaging the structures of soils during digging. There is a need to support finely textured soils, especially those capable of migration into coarse gravels. There is a real benefit to having a multiplicity of changes of interfaces of porous media, ever finer, through which effluents must pass before infiltrating soil surfaces. Perhaps the day will come when a man-made device will replace our historic use of heavy and cumbersome gravels.

SHORT-CIRCUITING

When soils are dug and replaced uncompacted, as they are when burying solid distribution pipes in trenches, those loose soils often provide avenues for effluents to flow where not wanted. Gareth Ott, a sanitarian in San Mateo County, California, has found it best to place a concrete dam in any backfilled trench conducting a solid distribution pipe where short-circuiting would be harmful. Concrete mixings are generally more readily available at a job site than is equipment for compacting soils.

COMMENTS

Codes through the years have rarely directed practices. It has been mostly the other way around. When certain constructions proved successful, codes were developed to demand an acceptable quality of performance. Once formalized, practices change only cautiously. Still, even the most dogmatic authority will tend to permit a change, provided responsibility rests with the entity proposing the change.

Contractors in many places would like to see changes in codes, and many authorities welcome cooperation with the local contracting industry. A united contracting industry prepared to document experiences could be a positive influence towards improvement of local septic-tank codes.

While speaking to the point of more involvement of a contractor with septic-tank practices, he is encouraged to obtain information on the growing trend towards ongoing management of septic-tank systems [46]. Septic-tank practices can be improved, and there are honest profits to be made by provision of maintenance services [98,99].

CHAPTER 9

ASSIMILATION

All of us have seen water flowing down nature's plumbing drains, swales, streams, creeks, and rivers, either to evaporate or to reach the ocean sumps. Our septic-tank effluents must somehow be returned to the natural water cycle. They must evaporate, be taken up and transpired by plants, become a part of a ground water basin, or reach the nearest stream. After all, there are no new waters on the face of the earth, and our wastewaters cannot simply disappear. They must go somewhere. Yet when designing a disposal field, many tend to view the soils of a site as a kind of gigantic sponge, intended to soak up everything put into it forever. Indeed, for the most part, the assimilation capacity of a residential site exceeds the needs (see Figure 25.)

Nationally provided guidelines almost entirely omit considerations of assimilation. Guidelines intended for certain purposes have been misunderstood, and misapplications have interfered with assimilation.

Not much quantitative information is had on assimilation, but what principles are known will be presented. Because of unfortunate misunderstandings of guidelines, some of the more troublesome will necessarily be discredited to provide the reasons local authorities need for improving their local practices.

In following discussions of assimilation, it seems best to use real experiences and pseudonyms to exemplify principles.

DARCY'S LAW

Mathematics are best avoided when presenting principles, but Darcy's law is a mathematical relationship [100]. So, statements of meaning will be presented to readers wishing to ignore algebra and trigonometry.

Figure 25. Not all men choose to herd with their fellows. Wastewater flows from comfortably spaced residences are easily assimilated. The capacity of high-density living areas to assimilate wastewaters cannot meet needs. Off-site assimilative capacities must be taxed.

Darcy's law essentially states that if a hillside is steeper or soils more permeable, subsurface waters will percolate downslope faster. And, twice as much water can percolate through twice as much cross-sectional soil area.

For our purposes, Darcy's law states:

$$Q = k\ i\ A$$

where Q is the amount of water percolating down a hillside through a wafer of topsoil with an area A being cross-sectional to the pathway of flow; k is the coefficient of permeability of the porous medium, the saturated soil; and, i is the hydraulic gradient, being the slope of the hill.

Q IS DIRECTLY PROPORTIONAL TO A—IMPLICATIONS THEREOF

Twice as much water can be percolated through two columns of soil as just one. That is, Q is directly proportional to A.

Consider a flatland where 3 ft of permeable soil overlaid an impervious substratum (glacial tills of northlands can be that way). Mutt and Jeff were neighbors and each wanted to build a house with a septic-tank system. The local codes demanded 4 ft of soil between the bottom of a disposal trench and an impervious substratum. Mutt and Jeff did not have that soil depth, but the sanitarian permitted mounds. A mound is a raised bed of imported fill, laid over native soils [101].

Mutt's mound was constructed 50 × 50 ft square in his backyard (looking much as though an elephant had been buried there). That mound had 2500 ft^2 of overall horizontal area. Effluents percolating downwards were stopped by the impervious substratum. They then percolated sideways out from under the perimeter of the mound. They then percolated further away in an ever radiating way to become assimilated by the surrounding area. Soils under the perimeter of Mutt's mound constituted the smallest cross-sectional area through which percolates needed to have passed. That is, Mutt's daily wastewater flow, Q, had to pass through a 3-ft-thick wafer of soil offering an area cross-sectional to the lateral pathway of flow amounting to: A = 3 × (50 + 50 + 50 + 50) = 600 $ft.^2$

Jeff built a same-sized mound of 2500 ft^2 of overall horizontal area. But Jeff wrapped his mound around the front and the side of his house to give aesthetic relief to his flat land. Altogether, Jeff's mound was 10 ft wide and 250 ft long. As with Mutt's mound, the effluents had to percolate sideways through soils 3 ft deep, and away from the perimeter of Jeff's long and narrow mound. There, however, the cross-sectional area was: A = 3 × (10 + 10 + 250 + 250) = 1,560 $ft.^2$

Assume that 1 ft^2 of cross-sectional area of soils of that area could percolate laterally only 0.4 gal/day (gpd). Mutt's mound could manage: 600 × 0.4 = 240 gpd. That compact, square mound caused troubles until Mutt's household learned to cut water usage from an average of 255 gpd down to 240 gpd. Jeff's long and narrow mound, on the other hand, could manage: 1560 × 0.4 = 624 gpd. Jeff's mound has not been troublesome.

Two Disposal Trench Systems

Suppose Mutt and Jeff had a neighbor with a trouble-free disposal trench system of the ordinary sort. Facing that physical reality, Mutt and Jeff decided to ignore the local guidelines demanding 4 ft of soil between the bottom of a disposal field and an impervious layer. Instead of having constructed mounds, at night each dug a disposal trench system with trenches 3 ft deep and with gravel fillings 2 ft deep.

Mutt believed that a system of four parallel trenches would serve. He dug each trench 1.5 ft wide, 50 ft long, and he spaced them at 6 ft on centers. Effluents from Mutt's septic tank infiltrated the soils through the sidewalls, but they could not percolate downwards or between trenches to get away, except out between trench ends. They could only be assimilated in the general area by percolating away from the outer periphery of the entire system. For simplification, assume the ends of the overall system of trenches had a periphery each of 20 ft. The entire system periphery was: 50 + 50 + 20 + 20 = 140 lineal feet. The cross-sectional area through which flow necessarily passed to get away in Mutt's 2 ft of useful sidewall height had been: $A = 2 \times 140 = 280\ \text{ft.}^2$

Jeff, however, decided to dig as much as Mutt did, but Jeff dug one long trench at the side of his property. His trench was 1.5 ft wide and 200 ft long. Jeff's septic-tank effluents were assimilated on both sides of his trench and at ends. Altogether, the cross-sectional area through which lateral percolation away from Jeff's 2 ft of useful sidewall height amounted to: $A = 2 \times (200 + 200 + 1.5 + 1.5) = 806\ \text{ft.}^2$

Again assume the soils could transmit lateral percolation at 0.4 gal/ft^2/day. The surrounding soils of Mutt's somewhat compact trench system could assimilate only: $280 \times 0.4 = 112$ gpd. Mutt's family, using 255 gpd needed help. On the other hand, soils around Jeff's long trench could assimilate: $806 \times 0.4 \simeq 322$ gpd. His 255-gpd family was served trouble-free.

For mathematical simplifications, examples assumed an impervious substratum which limited considerations to one plane. Rainfall was also ignored. The same principles would have been demonstrated if more complex mathematics had been applied to a deeply permeable soil, that is, a three-dimensional model.

Inferences

In flat terrain, assimilation of percolating effluents is assisted by providing a disposal field with as much perimeter as practical. Long and narrow rather than compact arrangements are best.

When permeable soils in flat terrain are overlaying a practically impervious substratum (or perhaps ground water), a mound offers little if any advantage over a properly designed disposal trench system. This thought assumes no more damage of natural soils from digging disposal trenches than from constructing mounds.

In Passing

The writer has had only two occasions to recommend a kind of a mound. Both were to satisfy a local soil depth requirement rather than because of technical need.

In one of California's bays, for example, local authorities demanded 4 ft of soil below a bottom of a disposal field, and they would not relent because of no-growth activists. Four feet happened to be the total depth of soil on Bill's lot to ground waters. At the same time, activists had managed to enact an ordinance prohibiting engineered fills. No fill soils could be imported for a mound. Bill looked for help and was rescued. He now has one of the world's largest "bottomless flower boxes" containing a subsurface reclamation and irrigation system.

SLOPES AND HYDRAULIC GRADIENT

The hydraulic gradient of water percolating downslope in a permeable soil stratum overlaying a less permeable stratum increases as the slope increases. That is, water percolates down a steep slope faster than it does a slight one. The hydraulic gradient of water percolating straight downwards is unity. In a horizontal column of soil, it is zero.

Mathematically, the gradient is equal to the sine of the angle of the slope. The sine and the tangent (slope) of small angles are about equal. So, the slope up to 16% could be used for the gradient with two-place accuracy. Slopes up to 34% can be used for gradient values up to one-place accuracy.

At Chigger camp, a few years ago, a septic-tank system was needed to manage 12,000 gpd. A hillside sloped at 58% was the only candidate area. A USPHS percolation test rate of 10 min/in. was descriptive. Soils were 12 to 15 ft deep over weathered granite.

The local sanitarian wanted disposal trenches no longer than 100 ft long each, and no deeper than 8 ft. The *Manual of Septic-Tank Practice*—local practice—advised 1.6 gal/ft^2/day for sidewalls of pits, and the 8-ft-deep trench was considered a pit. The engineer designed five trenches, 100 ft long each, plus a trench 36 ft long, for a total of 536 ft. One trench was to be built above the other towards the hilltop where the slope lessened and a backhoe could work without first cutting terraces. The engineer also wanted as much of the slope between his disposal field trenches and a stream below as possible.

Another consultant to Chigger camp pointed out need for consideration of assimilation, a new thought at the time. The slope below the stacked trenches essentially offered a percolation swath 100 ft wide and 12 ft deep at most: A = 1200 ft.^{2} Assuming complete saturation of the upper soil layer and applying Darcy's law, Q was 12,000 gpd, or about 1600 ft^3/day. The cross-sectional area, A, through which flow was to occur was 1200 ft.^{2} The steep slope was about 58% (30°), and the sine of that angle being the hydraulic gradient yielded: i = 0.5. Solving for the coefficient of permeability yielded: $k = 1600 \text{ ft}^3 \div 1200 \text{ ft}^2 \div 0.5 \simeq 2.7$ ft/day, or 9 ×

10^{-4}cm/sec. Then the formula for translating such coefficients of permeability into estimates of USPHS percolation rates was applied (Chapter 3). Using log p = 3.064 × 0.644 log k yielded 4.7 in./hr, or about 13 min/in. Inasmuch as the design percolation rate was 10 min/in. and complete saturation of the slope's soils had been assumed, the engineer and sanitarian were uncomfortable and relented. More yet, precipitation had been left out of considerations.

The next design had one trench with 10 ft of useful sidewall height for infiltration. That trench was 375 ft long along one contour. The swath of downslope percolation could be through a completely saturated 12-ft-thick soil layer: A = 4500 $ft.^2$ Again substituting into Darcy's law and solving for the coefficient of permeability yielded: k = 0.71 ft/day, or 2.5 × 10^{-4}cm/sec. That corresponded to a percolation rate of about 5.6 in./hr, or 10.8 min/in. It was uncomfortably close, so the engineer finally lengthened the trench to 400 ft and hoped the granite was fractured enough to help. It was.

Inference

Disposal trenches on hillsides, especially where a less permeable substratum underlays a more permeable one, should consist of a single trench (if possible) occupying a single, uppermost available contour.

Comments

Application of Darcy's law to hillside situations has been found satisfactory with the simplifying assumption of saturation of the pathway of flow, the swath being as long as the downside edge of the trench system, as deep as the permeable soil layer, and a soil being consistently permeable throughout its mass. That last assumption, together with using the formula for translating coefficients of permeability into percolation test rates, and vice versa, are usually the least well known conditions. Therefore, solving for k and p (coefficient of permeability and percolation rate in in./hr,) respectively seems the wisest approach, and then working designs to the safe side of those values.

Flow nets also constitute an approach to assimilation problems [6]. They are more time-consuming, and they are especially time-consuming in trial and error adjustments. And, they require at least as questionable assumptions as does the method presented herein.

Local authorities would be best advised to insist on estimates of assimilative capacity for rather large disposal fields. Such projects should be funded for adequate site investigations and competent consultants. As for standard individual disposal fields, authorities must follow guidelines, unless for some reason special studies are made. Those guidelines, nevertheless, should not result in inferior systems. Unfortunately, they sometimes do.

A FALLACIOUS GUIDELINE: MAXIMUM LENGTH OF DISPOSAL TRENCHES IS TO BE 100 FEET

Perusal of the *Manual of Septic-Tank Practice* will disclose a strong tendency for our Eastern-situated national authorities to look to flatlands, much as this writer, a Westerner, tends to think in terms of mountainsides. Thus the authors of the *Manual* apparently thought of hydraulic gradients as either being unity—straight downwards forever—or a frustrated zero when percolating effluents encounter bedrock or a water table too close for comfort. They had the tools to consider using Darcy's law (known to engineers since 1856) for development of principles, but the needs were not obvious.

If one views the drawings of a "typical layout of absorption trench," in the *Manual*, he will see three parallel disposal trenches joined at their ends such as to form hydraulically a single trench with changes of direction at corners. Occasionally, traffic of heavy machinery had been known to break rather brittle clay tiles in trenches. That caused authors of the 1957 edition of the *Manual* [92] to advise, "The individual laterals preferably should not be over 60 feet long, with a maximum length of 100 feet, and the trench bottom and tile distribution lines should be laid at a grade of 2 to 4 inches per 100 feet and never to exceed 6 inches per 100 feet. . . . Use of more and shorter laterals is preferred because if something should happen to one line, most of the field will still be serviceable." Perforated plastic pipes did not reach the market until about 1972. So brittle tiles were still in vogue when the 1967 revised *Manual* [15] advised, "The individual laterals should not be over 100 feet long, and the trench bottom and tile distribution lines should be level." The same reason of fear of disturbing one line was offered.

Authors of the *Manual* failed to notice that disruption of either the pipe from the septic tank or all laterals would be needed to much damage the continuous, flatland system they advised. Each lateral could be served effluent from either end. Further, effluents could easily bypass a broken tile through gravel fillings.

Nevertheless, many local authorities took the *Manual*'s assertions at face value and forbade distribution trenches to be over 100 ft long. The

UPC author committee did also. The consequences have been disastrous. Many homebuilders have had severe inconveniences in fitting trenches onto their lots. Many fields on slopes consist of tiers of trenches in shallowly permeable soils where about all effluents from the highest trenches percolate through the trenches below, adding to their loads.

Suggestions

On hillsides, utilize a single contour for trenches of whatever length satisfies infiltration surface requirements and assimilation needs. Avoid (but not summarily prohibit) one trench above another on different contours. Either keep heavy machinery off the disposal field area, or provide distribution pipes strong enough to withstand heavy traffic. Also, locate the disposal field contour as high on a slope above a roadway or drainage swale as possible, but not necessarily such as to interfere with gravity flow arrangements from the house. Whatever deep percolation and evapotranspiration that hillside can offer to assimilating effluents percolating downslope should be utilized.

Although the *Manual of Septic-Tank Practice* and the *Uniform Plumbing Code* provide for local judgments overriding detailed provisions, sometimes a local authority will not use his assigned authority to vary a detailed guideline. Still, most guidelines pertinent to maximum trench length permit a same accomplishment as advised above, trenches using more than 100 ft of a single contour. The designer, for example, could have several trenches all on a single contour. He could have 100-ft-long sections of trenches with ends separated by 6 ft of an undisturbed soil berm (generally 6 ft is the required spacing between trenches). Through that berm, a solid distribution pipe could connect the trenches in tandem. A firm, but prayer-rug position might also be advised to the designer, because some local authorities rather dislike situations betraying illogics of their codes.

A FALLACIOUS GUIDELINE: FOUR FEET OF SOIL BELOW THE DISPOSAL FIELD BOTTOM

During the FHA-sponsored USPHS field investigations, troubled homes and disposal fields were occasionally found in subsurface drainage swales or topographical basins. Obviously, local practices failed to identify such problem spots, so it was hoped a guideline would make up for lack of competent in-field judgments [21]. The authors of the *Manual* therefore advised, "The maximum seasonal elevation of the ground water table should be at least 4-feet below the bottom of the trench or seepage pit. . . . Rock formations or other impervious strata should be at a depth greater than 4-feet below the bottom of trench or seepage pit."

Whenever a guideline fails directly to identify a problem and the authors hope to accomplish something indirectly, the outcome too often is a failure to identify the situation of concern and senseless application of the guidelines to situations not of concern. There is a need for local authorities to recognize drainage swales, topographical basins, and other soil and geological aspects where neither a house nor a disposal field belongs.

Laak provided a method of estimating the needed depth to ground waters or impervious rock strata [6]. Perusal of his approach will disclose the hopelessness of a hapless, universally advised number of feet to be below disposal field bottoms.

As a matter of pertinent interest, seasonal ground waters on hillsides are often not observed during the extended dry season of California. Unintentionally, disposal trenches are often constructed, the lower portions (or all) of which are immersed when seasonal rains come and cause a downslope percolating, perched ground water table. But there happen to be thousands of such disposal fields serving properly. Infiltrating effluents successfully join the downslope percolating ground waters, and whatever health hazards as may be present are filtered out of percolating waters during their travels through soils.

Inasmuch as the USPHS did not wish to insult local authorities who could not recognize the need to stay out of drainage swales and basins, no reason for the 4-ft-of-soil guideline was offered. Most unfortunately some state authorities in the Southwest (perhaps elsewhere too) have presumed that the reason for that guideline was a fear of contaminating groundwaters. The ambiguous *Manual* could be so misunderstood. At the same time, those state authorities failed to notice that the *Manual* provided public-health-oriented setbacks of scores of feet from wells and streams. No one would likely permit a disposal field only 4 ft away from water supplies. As said, that misunderstanding has been most unfortunate. A 4-ft-of-soil guideline has been used to justify mounds and ET systems in too often unsuccessful replacement of theretofore satisfactory practices.

There are ridiculous extremes too. On one occasion, the writer was asked how to prevent wave action from beating in the top of a septic tank. That tank was exposed during low tide in the San Francisco Bay, but immersed during high tide. The breakers had damaged the tank. Judiciously, the disposition of the City Councilman's disposal field was not discussed.

In the arid West, small rivers often have huge, broad floodplains. The same as people build on volcano slopes (e.g., Mauna Loa and others), they also build in floodplains when years of quiescence lead to belief nothing will happen to upset matters. Eventually the volcano erupts, and eventually floods come. If people can survive such habitats, there seems to be no reason to forbid a disposal field in a floodplain. It would seem far better to have wastewaters under inundated soils than flowing out of flooded sewer manholes or out of disabled sewage treatment plants.

A FALLACIOUS GUIDELINE: TRENCHES MUST BE SPACED A MINIMUM DISTANCE APART

Spacing between elements of a disposal field is necessary for construction. Beyond that, only thoughts of assimilation could justify a particular spacing.

In satisfaction of a 300-ft^2 bottom area requirement, Joe decided to dig three trenches, each 1 ft wide and 100 ft long. His first two trenches were 2 ft apart on centers, 1 ft apart inside edges, and 3 ft apart on outer parallel edges. The local authority visited the site while Joe was working. That authority directed one of his two trenches to be backfilled because the local guidelines specified spacing to be 6 ft on centers.

Joe solved his problem. He dug out the 1-ft-wide berm between the two trenches. He then had a trench 3 ft wide, 100 ft long, and with 300 ft^2 of bottom area.

With spacing violating the local guideline, had Joe dug his three trenches as originally planned, he would have had 300 ft^2 of bottom surfaces plus 600 lineal ft of valuable sidewalls. The inferior, 3-ft-wide guideline trench had the same bottom area, but only 200 lineal ft of sidewalls (ignoring ends).

As was said, guidelines failing to deal directly with intentions often find senseless applications.

A FALLACIOUS GUIDELINE: TRENCH BOTTOM SHOULD BE LEVEL

Assume a disposal trench 1 ft wide, with 1 ft of useful sidewall height, and 50 ft long. Imagine that the trench had been installed at a sharp angle downwards on a hillside. Effluents would back up the trench until hydraulic head forced them to ground surfaces at the lower portion of the trench. That situation will not answer.

Next imagine that same downsloping trench had been dug in level ground with the top end below grade and the bottom end far below grade. Now that sloping trench would not have effluents surfacing, but it violates advice of the *Manual* that ". . . trench bottom and tile distribution line should be level."

Finally, imagine the trench is vertical in the ground. Now it is a pit 1 ft square and 50 ft deep. It meets guidelines now that the violation of the directive to be level is so extreme as to be vertical.

A sloping bottom of a disposal field element indeed can serve. Soil surfaces which could be inundated by effluents should be considered in satisfaction of infiltration surface requirements (sidewalls being considered too, rather than bottom surfaces only). Also, it is necessary that arrangements will not result in surfacing effluents before all useful infiltrative surfaces are utilized.

A strongly sloping bottom might cause difficulty with local guidelines illogically distinguishing bottoms of trenches from sidewalls of pits in estimating infiltration surfaces. And, the geometry of very irregular excavations is an inconvenience to an inspector calculating requirements.

At times, belief in the need for a straight and level trench bottom has resulted in blasting a rock out of the way with dynamite. Trenches need not be straight, unless on a contour of a hillside. There, a disposal trench can tolerate a rock blocking a place as easily as two trenches could serve with a solid pipe in a soil berm between them. It is only necessary to serve each sector of a trench with a distribution pipe bypassing a rock or other obstruction. The obstructed trench could be regarded as two trenches with serial distribution.

A FALLACIOUS GUIDELINE: A HORIZONTAL SETBACK ON A SLOPE

Eastern geology varies, but one tends to think of rather flat sedimentary deposits as commonplace. When a flat bedrock is not much fractured and it outcrops to daylight on a hillside, a disposal field above could easily have effluents percolating to daylight below. The slope of a hillside, as such, would seem inconsequential. The primary contributing factors would be closeness of the disposal field to the rock and the surface of its outcropping.

To be sure, western geology has sedimentary deposits. But, the thought is more commonly of rugged terrain with percolating rainwaters seeping from hillsides with roadcuts intersecting less permeable substrata running almost parallel to downsloping surface topographies. A disposal field just above such a roadcut would likely yield percolating effluents to daylight below. Enough of those situations have occurred to cause authoritatively provided guidelines with hopes of avoiding such problems.

Unfortunately, guidelines have been fallaciously directed towards slopes, as such, and have not been directed towards potential problem situations. Consider the 1976 Edition of the *Uniform Plumbing Code* [47]. Appended to a table as a footnote, one finds, "When disposal fields and/or seepage pits are installed in sloping ground the minimum horizontal distance between any part of the leaching system and ground surface shall be fifteen (15) feet."

Consider an unfractured sedimentary rock slightly sloping and outcropping on a hillside. On the 20% sloping hillside, the disposal field could be placed about 25 ft away from the outcropping below. But, the soils are deeply permeable and the effluent percolates to the rock which inclines slightly towards the outcropping. Effluents will percolate to daylight.

One of the most common causes of effluents percolating to daylight in hilly terrain comes from the cutting of a roadway to a garage. The roadway will stand a good chance of intersecting a less permeable substratum over which ground waters tend to percolate. An experienced sanitarian might foresee a disposal field being too close to a chance roadcut which may be constructed after the disposal field is constructed and inspections have been final. But a less experienced sanitarian will not foresee the possibility.

Guidelines failing to deal directly with intentions often fail to solve problems. There is no substitute for identifying a potential problem and requesting a technician to solve it.

REALITY TESTS

Mother Nature is not neat. Despite percolation tests, explorations of backhoe holes, and opinions of soil scientists and geologists, no one knows the subsurface characteristics of a site without impractical numbers of investigations. Sometimes it is needed to know with certainty that the assimilative capacity of a site is adequate for intended purposes. Simulations of real situations are far more convincing than lesser testing programs.

Assimilative Capacity at Rocky Acres

A substratum of a sandy soil was candidate for disposal of collected wastewaters from 23 planned, individual household lots. Although about

6000 gpd were actually expected wastewater flows, county guidelines mandated 8000 gpd in design estimates.

The only likely area for injection was a small basin. The soil was so filled with rocks larger than boulders that customary disposal fields would have been impossible to construct.

It was decided that gravel could be used to form a level surface within the basin, and an almost maintenance-free sand filter [11] could be placed over the gravel. Only the assimilative capacity of the basin was in question. Estimates could not be obtained because deep explorations of the soils were impossible because of huge rocks. Even a percolation test hole could not be dug down between the rocks. A reality test was needed. Water service to the site had been constructed in anticipation of homebuilding, and a fire hydrant was nearby. It was opened with a fire hose discharging into the basin.

During a 24-hr period, an area within the basin amounting to about 2000 ft^2 accepted about 237,000 gallons of fresh water. Inasmuch as 8000 gpd was the design value and a nonclogging sand-filter effluent was contemplated, the assimilative capacity of the site was known to be, at the very least, 30 times what was needed.

Assimilative Capacity at Eagle's Nest

A long-ago approved lot could not meet modern codes. The builder constructed a foundation, but was then stopped by the local authority on a needed septic-tank permit. The lot had a building pad cut on a very steep hillside. The outside edge of that particular pad was a cliff displaying fractured bedrocks. Immediately below the pad was another building pad belonging to the neighbor's lot.

The subject pad would accommodate a disposal trench system, but the authority questioned whether or not effluents would surface through the fractured rock cliff and run onto the pad below.

The builder was instructed to construct a disposal trench in the area where the field was to be located. He was to waterfill that trench daily. At the end of about a couple of weeks, the local authority would then know with reasonable surety whether or not effluents from a septic-tank system would reach daylight at the cliff's edge.

Assimilative Capacity of Vertical Estates

When a California state authority was convinced that steep hillsides could not serve for disposal trenches without bringing effluents to daylight, a distressed subdivider supported tests pertinent to the point. Findings on Lot A were typical.

On Lot A, slope was 70%. Soils were about 2 to 3 ft thick over soil-filled, fractured rocks. A trench was hand-dug with pick and shovel along a contour 24 ft long, 14 in. wide, and 18 in. deep. Pea gravel was placed 1 ft deep into the trench by sliding it down a chute from above (an easier task than first expected), and a perforated distribution pipe had been embedded within the gravel. That pipe had been fitted with a riser pipe to above backfilled soils. Daily, 55 gal of fresh water were discharged into the riser pipe from a water-tank truck.

Each addition of water was observed through the riser pipe to infiltrate at 3 min/in. It was obvious that the trench could have been water-filled several times daily, rather than on the once-a-day schedule chosen.

At the end of 29 days, the trench was cut open on its downhill side. Wet soils indicated pathways of percolation. Wet soils displayed no tendency for waters to surface, rather they percolated a few feet downslope before seeping into the fractured rock stratum. Both the soils and rock stratum had permeabilities underestimated by percolation tests.

That reality test on Lot A demonstrated the practicality of hand-dug disposal trenches (authoritatively questioned in this modern world of machinery), and that a slope almost too steep to walk on could accept a disposal trench of proper design. The authority relented on slope, 4 ft of soil below the bottom of a disposal field, and similar requirements in view of the indisputable reality test.

The assimilative capacity of the demonstration trench was doubtless far greater than demonstrated. Under the test conditions imposed, however, that trench displayed an unquestioned downslope assimilative capacity of about 2.3 gpd per lineal foot. Even if a water-wasting family discharged 600 gpd, only 262 lineal feet of such a trench would be needed along a single contour. The assimilative capacity of the site far exceeded ability to inject wastewaters in disposal trenches.

ASSIMILATION BY EVAPOTRANSPIRATION

Evaporation from soil surfaces and transpiration by plants (evaporation from leaves) have together become known as *evapotranspiration* (ET). Evaporation and transpiration are surface area phenomena. The total amount of water evapotranspired is a matter of how fast it evapotranspires from a surface and how much surface there is. Thus, considering ET for assimilation of wastewaters is always a matter of getting wastewaters to evapotranspiring surfaces and providing as much surface area as possible.

About two decades ago, ET was a process of rare promotion and then promoted by incredible claims. Most authorities neither understood ET nor believed it could be used to dispose of wastewaters. Pioneering efforts of Alfred P. Bernhart at the University of Toronto should be credited for ET having achieved widespread attention. Authorities no longer use the question, "Do you believe in ET?" as subject of casual business conversation.

As is often observed, underdone goes to overdone. Earlier beliefs that ET cannot work have gone to current beliefs that ET can always work. Consequently, it is often used in poor designs, poor constructions, and poor locations. The results have too often been unhappy. One director of environmental health, for example, accepted a subdivision with about 200 lots onto which ET systems were to go. After the first six systems failed, the director began job seeking. Similar stories are legion.

ET *can* work, but whether it *will* work is another question. ET is not a simple process to quantify and to predict. Experts make a life's study of ET in agricultural pursuits. Wastewater disposal systems are at least as complex and they have received far less scientific attention. Therefore, wastewater management systems dependent mostly on ET cannot be designed with confidence-inspiring criteria.

Energy and ET

The evaporation of water is an energy-consuming process. It takes a while to bring almost freezing water to boiling, but it takes a far longer time on the stove to boil it away. Far more energy is required to evaporate water than to warm it from almost freezing to boiling.

The energy for ET comes directly and indirectly from sunlight. Clothes on a line dry faster in sunshine than shade. But the sun's energy expressed in drying winds—advection—also will power ET. Even in the shade, clothes on the line will dry fast on a windy day.

Potential ET is that which would be achieved if the water in soil and plants was unlimited. That is, there is much potential ET in the dry and hot American desert, but little ET actually occurs. It is an objective of a designer of a wastewater ET system to utilize all of the potential ET there is at a subject site.

Potential ET and evaporation from lakes are practically alike. Designers of ET systems can find help from scientists of the U.S. Weather Bureau. Kohler et al. [102] were interested in using records of Class A pan evaporation to predict annual lake evaporation. Their studies yielded maps of the United States giving isopleths (like contour lines) of "Average Annual Class A Pan Evaporation in Inches," and "Average Annual Lake Evaporation in Inches." Because evaporation from pans is greater than from comparatively much larger surfaces, annual pan values must be converted into lake evaporation by a factor. So, Kohler et al. also produced, "Average Annual Class A Pan Coefficient in Percent." Those values of the ratio of lake evaporation to pan evaporation varied over the country from 60% at places in the American southwest desert to 81% at the extreme northeastern part of Maine.

As one would expect, potential ET is greatest in summer and least in winter. ET from a large lysimeter planted to grass turf at the University of California in Davis has been measured [103]. ET varied from a summer high of about 9 in./month to a winter low of about 1 in./month. Values varied yearly also.

The best indicator of ET is pan evaporation, but oftentimes it is not a routine measurement of weather stations. So scientists have provided formulas for using other weather station records and hydrologic reports to estimate regional ET [104,105].

Microclimate

Regional ET may be estimated. Within the region, however, the wastewater ET system will be situated within a microclimate. Quantifications cannot be had. Various site features affecting microclimate can only be understood in principles and subjectively managed.

The microclimate of a wastewater ET system is very much affected by man's structures and plantings. A house will have a northside where shade will much reduce ET (Fig 26). Snow persists in shade, and shaded plants transpire far less than those in sunlight. The amount of shade will even determine the species of garden plants to be successfully grown.

A tree can shade an ET system out of function. A designer of an ET system must caution the homeowner against tree plantings which could shade the system. Perhaps that homeowner may comply and instruct the next buyer of the house, and he in turn the next. Offering to manage the neighbor's gardening activities, however, may not be especially welcomed.

Advection, drying wind, is much affected by man's structures and gardening activities. Comments concerning control of a homeowner's and neighbor's decision to construct a garage, fence, or the like apply to thoughts of advection also. If the developed subject site is near a vacant lot, it is particularly difficult to entice a neighbor developing later to adjust his plans to the convenience of the existing ET system next door.

Figure 26. Even in freezing temperatures, sunlight melts snow and ice, but they persist in the shade. ET is very much less on the shaded northside of trees, houses, fences, and the like.

Effluent Storage and Accessibility

Where ET is looked to for the entire mode of wastewater assimilation, low ET rates must be taken into account. Except where money is practically unlimited, ET systems cannot be designed large enough to function on cold, overcast or rainy, winter days. And were they designed that huge, plants over the ET system would need supplementary irrigation at other times of the year. Ordinarily, total ET systems are designed with effluent storage capacities intended to endure periods of low ET.

Where storage is to be within a packing of gravel, wicks of sand or the like extending down into the gravel must be provided to transmit effluents upwards to overlaying surface soils. The capillary rise of water in gravel is miniscule. Without wicks of some sort, effluents in gravels cannot be expected to reach soils and plants above. Unfortunately, many advocates of total ET systems have recommended gravel storages without provisions for effluent transport upwards.

The designer of an ET system dependent on storage capacities should estimate: the storage needed as a budgetary consideration during ET; the rapidity with which a sand wick can supply effluents upwards to transpiring plants; the ability of the overhead soils to transport effluents laterally from the wicks to space wicks properly; and the construction needed to keep sand wicks in place. The designer will likely choose to very much overdesign, rather than trust questionable estimates. Still, he must know what overdesign is.

Oxygenation

Water uptake by plant roots is an oxygen-dependent metabolic process. If roots are in waters without dissolved oxygen, those waters will not be taken up. Consequently, unless wastewaters contain dissolved oxygen, they will not be prepared for plant transpiration in an ET system.

Septic-tank effluents rapidly dissolve oxygen as they percolate through porous media. Unless soils are excessively saturated, effluents therein will contain dissolved oxygen. Advocates of aerobics believe the dissolved oxygen is a plus. It is, unless the effluents stand ponded in storage. If their long-term biological oxygen demand is greater than their dissolved oxygen, and it most generally is, stored quiescent effluents from aerobics will soon have little or no dissolved oxygen. The designer of an ET system with much storage capacity must somehow account for oxygenation.

Trees

Trees transpire enormous quantities of water. An acre of mature oak trees can transpire 2000 to 2600 gpd [106]. A single mature maple tree in a 625-ft^2 lysimeter was observed to transpire 240 gpd [107]. Marsh [108] provided a table with both surface evaporation and daily water use expressed as a function of a tree's spread.

Sometimes plans for ET systems are drawn showing trees to which large amounts of ET are attributed. But, until the trees are grown and in place, the systems must be able to function without the trees.

Occasionally, a disposal field is to be constructed new or as a repair on a lot where landscaping exists. Where it is convenient, a disposal trench might advantageously be constructed alongside of trees. Care must be taken not to come so close as to damage the tree roots, and the distribution pipe must be well encased with gravel to discourage roots from entering the pipe and clogging it. In that respect, local knowledge of tree species is useful in knowing which trees would have roots likely to enter and clog pipes. Willows, for example, should be avoided. It also should be understood that evergreens transpire during wintertime, whereas deciduous trees have no transpiring leaves in winter.

ET Adjunct to Disposal Fields

It would seem that a conservative designer of a household wastewater management system might not choose to design systems depending on ET alone. There are too many unknown and uncontrollable variables to permit designing with confidence. In addition to the foregoing information which would leave a conscientious designer uncomfortable, there are moot issues dealing with rainfall infiltration, snow cover, salt accumulations, consequences of plant toxicities resulting from substances occasionally wasted to household drains, and others. Nevertheless, considerations of ET can be an adjunct to conservative designs of customary disposal fields.

Soils Less Permeable with Depth—Flatlands

Sometimes a soil is had which is reasonably permeable on top, but which becomes less permeable the deeper it goes. In such places, it is not unusual to find a disposal field which served for a short while unnoticed. Soon ponded effluents rose until surface soils became noticeably water-soaked by capillarity and plant roots took advantage of the water and minerals. Then the field served for quite some time by ET from a lush patch of weeds. That was unplanned ET.

Planned or unplanned, ET takes place everywhere. Unplanned, its activities oftentimes are displayed in a lawn wherein lays a system of disposal trenches. Especially when grass has been freshly cut, the trenches appear as parallel green stripes on a pale green background of unaffected grass. Some of the healthier green over the trenches is due to water, but most is due to minerals in septic-tank effluents.

Where slowly permeable soils become much less permeable with depth, deep trenches will mostly pond until some lateral percolation and ET are had. It would be better in such situations to use trenches shallower and mostly in the more permeable topsoils. The barriers should be permeable, and distribution pipes should be well within, or even towards the bottoms of gravel packings.

Much of ET from the shallow trench will be wider in overlaying soils than the trench below is wide. That is, ET could be regarded as being somewhat related to trench peripheries. As before with periphery-oriented phenomena, trenches should be narrower rather than wider. They also should not be spaced into a somewhat compact area.

Many homeowners object to greener grasses over a disposal trench, their not harmonizing with the remainder of a lawn and betraying a sewage-oriented device. That might be borne in mind when placing trenches. It might be possible to arrange them along garden borders or elsewhere where the greener color would not be objectionable.

Soils Less Permeable with Depth—Hillsides

On hillsides where a modestly permeable and shallow topsoil overlays a less permeable subsoil, and where those subsoils become less permeable with depth, assimilation by ET can help.

On such a hillside, place a shallow trench along a high contour. No more than absolutely necessary, trenches should not be stacked one above the other in a compact way. The effluents can percolate downslope and either infiltrate the poorly permeable subsoils on the way or become evapotranspired. Trees on the downslope can be very helpful with ET, deciduous trees being helpful in summer and evergreens being helpful the year around. Although trees may be placed such as to shade the hillside somewhat, if their roots are there and uptaking downslope percolating effluents, those trees might themselves evapotranspire more water than the smaller plants on the hillside might.

ET Intended Without Subsurface Assimilation

Often a seasonally high, perched ground water table is not so high as to interfere with a properly designed and constructed disposal field, but authorities become concerned from what are believed to be reasons of public health. In such places, some authorities have demanded ET systems with a watertight membrane at the bottom of the system.

In one of our less-populated southwestern states, Mr. Kredulus, a staff member of the State Health Department, was by an U.S. EPA grant sent north to take a course on small wastewater flows from academicians favored by EPA. Mr. Kredulus returned to try out his new wares.

In Victimsville, a high water table had been historically managed by local contractors, and systems had been generally successful. Victimsville became target for a new ET design with a membrane asserted to protect water supplies.

As things stand now, the sister agency, the State Building Department, responsible for enforcement, has had inspectors' lives threatened. The ET systems too frequently fail. That agency is out of patience with the State Health Department for having mandated ET and for failing to relent in that mandate. Before a Victimsville citizen can sell his home, his existing, functioning system must be excavated and replaced with an ET system with a poor chance of successful operation, if groundwater levels are higher than an arbitrarily set depth limit. The State Health Department is making all defensive efforts to justify its unpopular stand. Mr. Kredulus even argued that septic-tank effluents too close to ground waters pose a potential health hazard, while effluents from failing ET systems running into homes are nuisances. Most local contractors have learned to puncture the ET membranes when inspectors are absent, and that has helped. One or two contractors have disagreed with mandates, and revocation of their licenses is being pursued.

The story of mounds in that same state is less exciting, but that story is likewise less than pleasant.

It is indeed most unfortunate that good concepts cannot be immediately translated into field practices. The reality is they cannot. They must be cautiously tried in new locales and almost grow organically. That is a reality of the household sewage management world, appreciated by local people in the field, occasionally understood by regional and state authorities, but quite beyond comprehension of totally removed national authorities [109,110].

Designing household wastewater disposal systems to function mostly or entirely by ET would be a hazardous occupation. Simply "plugging in" a proprietary plan would be even more hazardous. The reader is advised to consider ET as an adjunct to a system expected to function on subsurface assimilation. As for someone determined to design a system to function by ET alone, he is on his own.

ASSIMILATION BY SURFACE DISCHARGE

In yesteryear, piping septic-tank effluents to the nearest watercourse was acceptable practice in rural America. There were not so many people then, and they were mimicking the big city habit of dumping raw or treated sewages into the nearest body of water. Then "dilution was the solution to pollution," and indeed, it oftentimes was. After all, pollution is simply the gathering together of generally innocuous substances into a large, concentrated, nocuous mess. Introducing wastes back into nature's assimilation system is the legitimate objective of sewage management, be it by soils, evaporation, or water.

As populations grew and man no longer looked to water courses as though they should be natural extensions of his sewers, the notion of household wastewaters discharging into watercourses became quite unpopular. The overreaction even had authorities concerned about superficial ground waters which had no beneficial use whatsoever.

Authoritative wishes to the contrary, it is at times necessary to treat household wastewaters as best as practical and then have them discharging to a ditch or watercourse nearby. Sand filters have been a most used device for such situations. Garden irrigation by laundry waters and by other such innocuous wastewaters has been less encouraged by authorities, but many homeowners resort to such means very commonly. During rains and winter snows, irrigation gives way to discharge into the nearest watercourse.

It is not practical to undergo considerations of the assimilative capacity of a body of water destined to receive wastewater flows from a household which has no choice but to discharge wastewaters as best as can be done. Still, there are times when water-saving devices can be afforded [111]. Also, Orlob's formula (Chapter 10 page 136) could be used to design a channel of sand which could be constructed for the renovation of wastewaters being percolated therethrough to a suitable point away from a house. In 1982, for example, Larry Baker designed a box containing a sand channel which, by changing directions in compartments, was quite long and would be expected to produce an innocuous effluent [112].

CHAPTER 10

PUBLIC HEALTH CONCERNS

Foibles of individuals vary, but taken together in society those foibles form rather consistent behavior patterns. When such general truths are expressed, humor is seen in matters far beyond anyone's power to do much to change things: witness *Parkinson's Law*, Parkinson's *The Law and The Profits*, *The Peter Principle*, and others. Parkinson pointed out the ability of people to focus on the trivial, while all but ignoring larger considerations. An inescapable truth, such has happened in societal concerns for health and safety. Few, if any, public budgets for attending to risks of health and safety could be shown to be proportional to incidence or magnitude of problems.

In wastewater management, concerns which authorities have had for discharges have been inversely proportional to volumes. They have also been inversely proportional to degree of treatment. Far more man-hours have been spent in measuring setbacks between disposal fields and wells than in separating people from exposure to their raw wastes.

RAW SEWAGES

In 1971, attention was drawn to Powell Slough [113]. That small impoundment of drainage waters was shared by the Colusa National Wildlife Refuge and a private duck hunting club in the Great Pacific Flyway. Through years of receiving inadequately settled sewages from a nearby city, Powell Slough had sludge deposits so extensive as to be undergoing methane fermentation.

Inasmuch as duck hunters used Powell Slough, the question of hazard to health was raised [114]. A knowledgeable authority of the California State Department of Public Health assured that sewage ponds were well known to be favorite places for duck hunting and the matter was not of much concern. People routinely waded in sewage ponds, shot ducks and other waterfowl in those ponds, ate their lunches there, and ate the wildfowl retrieved from the ponds. The matter might have offended the sensitivities of some, but the epidemiology was such that serious authoritative concerns could not be justified [115].

Public health authorities of California had occasion to investigate chances of contracting an infectious disease while working in a sewage treatment plant [116]. It was found that those workers who daily came into contact with sewages had health equal to that of the rest of California's population.

Despite precautions, sooner or later, analytical chemists working in wastewater laboratories accidentally pipette sewages into their mouths. Aside from injured sensitivities, nothing deleterious comes from the event or is expected.

SEPTIC-TANK TREATMENT OF MICROBES

Bacteria

Early engineers reporting performance of community septic tanks noted bacterial removals from wastewater flows varying approximately from 40 to 50% [117,118]. Still, removals of bacteria in septic tanks are only inferences from measurements of those going in and those going out. They may sometimes be, but very probably often are not, the same bacteria. It is the fate of specific pathogenic bacteria that is the concern, not changes in populations of harmless bacteria.

In general, pathogenic microbes are rather host specific and they fare badly outside of the host. It is not conceivable that most pathogenic organisms discharged from human hosts would find safety within a septic tank. The environment of the septic tank teems with great varieties of organisms, varying from miniscule viruses to insects and even to higher forms of life at times. Those adapted inhabitants of the septic tank would not likely be without predators to the incoming pathogens.

While studying the microbiology of contents of septic tanks, Calabro [119] studied survival of *Salmonella typhimurium*. Calabro noted that enteric pathogen to die out within 2 weeks, ". . . and does not appear to pose any immediate danger to the handlers of septages."

Viruses

Viruses are so small (0.01–0.45 μ) that electrical charges of their surfaces may determine their destinies.

Studies at the University of California, and later private studies, established beyond doubt that suspended solids in septic tanks were electrically charged. They responded to cationic and anionic flocculants, depending on whether the sewages were comparatively fresh or many years old [9,10]. Inasmuch as viruses are charged particles and respond to flocculants [120], it seems likely that viruses would become attached to septic-tank solids and share their fate in sedimentation.

Investigations reported by the U.S. Geological Survey included cooperative studies of the University of Miami [121,122]. It was found that septic-tank sludge particles exhibited polyelectrolytic (flocculant) behavior, and when viruses were added to wet sludge none appeared in the supernatant fluid. The viruses attached to the sludge could only be removed by increasing the reaction to pH 8.0, a condition not normally found in septic tanks.

A seeding study was undertaken with a known quantity of attentuated polio viruses which was injected into a septic tank. Samples collected from the septic-tank outlet and from observation wells in ground waters below disposal trenches contained no viruses.

FILTRATION

It would be expected that predators of parasitic worms, their eggs, and the like would attack those pathogenic organisms within a septic tank. Assuming any of those relatively large organisms survive the hostile environment of a septic tank, they would not be expected to infilter the finer pores of the customary soil disposal field. The actions of strainers and sieves need no elaboration. The organisms would be expected to remain in a disposal field until death overtakes them.

ENTRAPPING WITHIN SOILS

Bacteria and viruses are smaller than some soil pores and they could travel through them, but their movements are impeded. Krone [123] described the removal of pathogens by straining at the soil surfaces, straining by intergrain contacts, sedimentation, and sorption by soil particles.

Assuming viruses somehow survive the hostile environment of a septic tank (which they apparently do not), soil constitutes a likewise hostile environment. Even coarse sands, as well as more electrically active silts and clays, remove viruses from percolating waters by virtue of electrical attractions [124]. A virus has about as much chance of passing through soil pores as a miniscule particle of iron would in passing through an unimaginably huge pile of magnets.

Spherical, cylindrical, and helical bacteria, respectively, are about 0.5–1 μ, 0.5–3μ, and 0.5–15 μ in their dimensions. Silt particles are 50 to 2 μ. Clay particles are less than 2 μ. Thus, a bacterium might find soil pores passable only by virtue of soil structure forming pores larger than the textural particles themselves. Like viruses, bacteria have surface charges. But being larger than viruses, bacteria are less likely to have their fates so determinable by surface charges. Bacteria travel farther in soils than viruses do, but not indefinitely.

The movement of bacteria through porous media was studied at the University of California [125,126]. Movement and removals as related to percolation velocities and particle size of porous media were formulated. Later, one of those investigators provided a simplified formula for field application [127]. Orlob's formula states:

$$L = 107\ d\ V$$

where L = required distance in feet for underground travel necessary to reduce sewage bacteria concentrations to the level prescribed for drinking water
d = effective size of sand grains in the porous formation in centimeters
V = velocity of ground water movement in ft/day.

For example, if a soil happened to have about 10% of its particles finer than silt (a sand or loamy sand), then d = 0.005 cm. If percolation occurred at V = 10 ft/day—a very rapid rate—Orlob's formula would yield a setback of L = 5.3 ft. Generally, rather small values for L are calculated when dealing even with coarse sands. Nevertheless, questions of large pores in highly structured soils, worm holes, and other channels not reflected in a textural soil description arise. Also, the variables needed in estimates are generally unknown. The formula has not been widely accepted.

A PRACTICAL APPROACH TO SOIL SITUATIONS

During the later 1960s and into the 1970s, subdivision of land in Colorado occurred at alarming rates. Mountain homes, theretofore of essentially no concern with septic-tank systems and wells, aroused fears of widespread pollution of ground water supplies.

The State Board of Examiners of Water Well and Pump Installation Contractors requested guidance from the Ground Water Section of the Division of Water Resources. In response, John Romero, Assistant Water Resource Engineer, exhaustively gathered information on what was known on travel of bacteria and viruses through soils [128]. Following are his findings:

> 1. Bacteria and viruses travel with the flow of water; they do not travel or move against the current.
>
> 2. Bacteria and viruses can move in a direction opposite that of normal ground-water gradients during times of recharge and/or pumping. The latter is an assumption based on the characteristics of well hydraulics.
>
> 3. In general bacteria and viruses are removed by the aquifer in the same manner as are coliforms.
>
> 4. The rate of bacterial and virus removal with distance is a function of an aquifer characteristic termed 'filterability.'
>
> 5. For any degree of filterability it depends on distance only and NOT UPON THE RATE OF POLLUTANT RECHARGE.
>
> 6. Aquifer materials best suited for the removal of biological contaminants are those uniformly composed of very fine to fine grained sand with a high clay content.
>
> 7. For an ideal system the maximum length of travel of biological pollutants with ground water ranges between 50 and 100 feet.
>
> 8. Pollution travel in nonsaturated systems is considerably less than that in saturated systems in that maximum lengths of travel appear to be in the vicinity of 10 feet.
>
> 9. The nature of the soil in contact with the source of contamination plays a dominant role in the subsequent travel of bacteria.
>
> 10. Bacteria and/or virus-infested pollutants might travel much farther than predicted if nutrient-laden waters are intercepted during the course of penetration.
>
> 11. Under favorable conditions (not necessarily applicable to Colorado) bacteria and viruses have been known to survive up to five years.

Romero provided double-line graphs, Figures 27 and 28, which delineated three zones. The "Prohibitive" and "Probably Safe" zones enclosed the "Hazardous" zone. The hazardous zone included well-documented case histories of biological pollutants travelling within media of a known particle size (Hazen's effective size, that being the size which 10% of the medium is finer than).

According to the empirical relationships expressed in Romero's graphs derived from case histories, any soil containing 10% silt or clay—both particles being smaller than 0.05 mm—would probably be safe at very

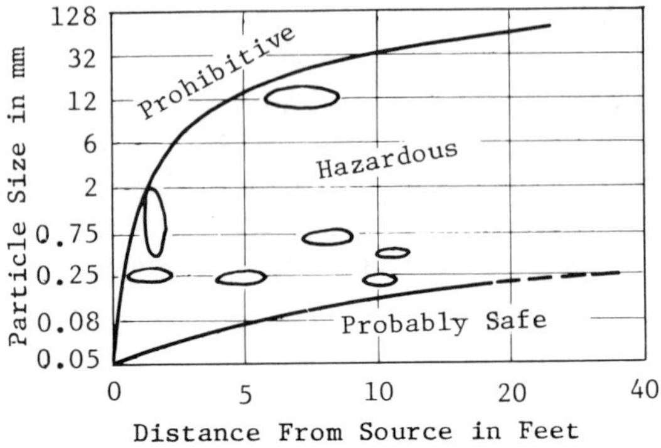

Figure 27. Biological pollution travel in nonsaturated porous media [128].

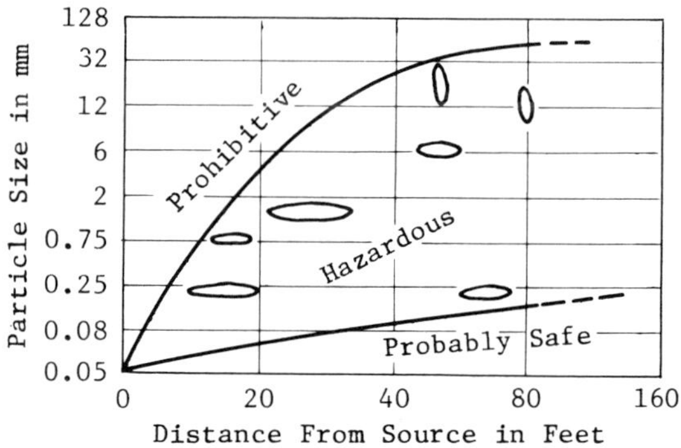

Figure 28. Biological pollution travel in saturated media, with ground-waters. Zones of maximum travel of biological pollutants are as described in documented case histories [128].

small distances between a disposal field and a fresh water supply. Such soils are classified as sands by the USDA Soil Conservation Service. Loamy sands, loams, and other such textural classes all contain more than 10% silt or clay. The range of 0.25–0.05 mm given in Romero's graphs is the classification of fine sand, and very fine sand. The range of 2.0–0.25 mm covers the range of very coarse sand, coarse sand, and medium sand. Anything coarser than 2 mm is no longer classified as soil [37,129]. Pebbles are sized from 4 to 64 mm, and cobbles are sized from 64 to 256 mm.

One could imagine a bacterium being swept-along with a percolation stream in a water-filled soil pore. Entrapment at a point of contact between two particles might not happen for a distance in that saturated soil, because the bacterium might be in the center of the percolation stream within the pore. But in unsaturated soils, waters and their swept-along bacteria travel through smaller pores and in films sliding over soil particles. Entrapment would be sooner. And as Romero's graphs show, the setbacks from the source of pollution and fresh water supplies is four times less in unsaturated soils. Nevertheless, even if a biological pollutant is injected directly into ground waters in what most people would recognize as soils, removal of that pollutant is extremely rapid from percolating waters in those saturated media.

Faced with the responsibility of offering recommendations to people already fearful of widespread contamination of ground water supplies, Romero recommended essentially the overly conservative setbacks in vogue. The USPHS *Manual* [15] typifies such recommendations.

Romero's graphs have found application, at least in California, where subdivision situations justify the expense of on-site studies needed for proper use of Romero's data. Despite the small setbacks one may derive from studies, authorities have found Romero's empirical case study approach appealing. It constituted a rational approach to setbacks, theretofore with no rational foundations for being.

Romero's collected data and graphs demonstrated the improbability of septic-tank systems in soils having polluted Colorado's ground waters. Oddly, it did not occur to anyone to question either sampling techniques and laboratory analyses or the public health significance (or lack thereof) of the 1 coliform/100 ml standard of maximum tolerance in Colorado. Laboratory technicians come in qualities rather than only in perfections. And, the generally good public health in Colorado did not conform with the fears of pollution.

GEOLOGICAL APPROACHES

Having collected data essentially vindicating septic-tank systems in soils as suspects of polluting ground waters, Romero turned suspicions to coarser porous media. He noted, "Areas underlain by igneous, metamorphic, or consolidated sedimentary rocks should be given a more critical examination because of the danger of extensive pollution travel via fractures and solution cavities."

Geological aspects of pollution travel in fractured rocks were investigated at Colorado State University. Geological aspects of pollution travel in solution cavities have been investigated in Dade County (wherein lies Miami), Florida, by the U.S. Geological Survey.

Fractured Granite and Grus—Millon's Studies

In 1970, Millon used total coliform counts as indicators of septic-tank effluents having reached well waters [130]. Fluorescein tracing had failed. Total coliform counts included fecal coliforms plus the ordinary coliforms which are usually inhabitants of soils.

In the Red Feather Lakes area, in Colorado, there were 700 families of fecal–coliform-producing people. In addition, there were the ordinary coliform inhabitants of soils. The number of total coliforms possibly available for the travel into shallow ground waters was unfathomably great.

Daily, each person discharges from 100 to 400 billion fecal coliforms [4]. As a point of reference, daily each person produces far more fecal coliform organisms than there have been total people on the face of the earth since the beginning of time. Or, if each coliform organism was as large as an ordinary grain of sand, a string of them would wrap around the earth's equator more than twice.

The Red Feather Lakes area was a small, mostly recreational development situated over medium-grained, faulted and jointed, Silver Plume-granite. Superficially, there were 3–4 ft of grus which graded into less well decomposed granite from 3 to 5 ft of depth. Below was somewhat weathered, but for the most part quite fresh granite.

Grus describes angular, coarse-grained fragments resulting from the granular disintegration of crystalline rocks. At the ground surface, disturbed grus has appearances of the sand and gravel it is. Undisturbed below, angular particles are fitted together as in the original rock, and a roadcut in such grus will display a distinct reminiscence of the former bedrock (Figure 29). Whereas soils commonly have about 50% porosity, an upper limit of porosity of in-place grus would be about 5%. Rocks fractured as they were at the Colorado study site would likely be less than 1% in overall porosity [131].

Ground waters were shallow. In the 85 shallow wells studied, ground waters averaged about 13 ft below ground surface and varied from zero to 37 ft deep. Some wells had been drilled to 80 ft deep (well below ground water levels), but most had been hand-dug to an average of 20 ft deep. The direction of ground water flow was governed by the faults and joints in the granite.

Bacterial counts did not display consistent patterns. Sometimes a well would have 300 coliforms/100 ml, and no coliforms 2 days later. About half of the analytical results were zero, 300 coliforms/100 ml was a common high, and 500 coliforms/100 ml was measured on two occasions.

The study area was mapped as consisting of about 7.6 $mi.^2$ It was estimated that 700 families inhabited the area before Labor Day. That

Figure 29. A roadcut displaying grus, the crystalline decomposition of granite, showing reminiscences of the original rock. Permeabilities and pathways of flow are as erratic as the weathering pattern.

amounted to a density of about 84,000 ft^2/person, assuming the average family at 3.6 persons. During the summer occupancy of the August study period alone, each person contributed about 1.25×10^{11} fecal coliforms to flows of septic-tank effluents. That assumes half of the fecal coliforms do not survive their stay in septic tanks [117,118,132]. Few of those effluents would have failed readily to reach the shallow ground waters which according to Millon's descriptions were largely in fractured granites rather than grus. As following studies will disclose, movement of pollutants in ground waters in fractured rocks and solution cavities is largely lateral, with very little downwards mixing. Likely, 20 ft of ground water depth would have been the lowermost limit of any mixing. Given those assumptions and estimating 1% porosity, about 17,000 ft^3 of water would have been contained within each person's 84,000 ft^2 of aquifer 20 ft thick. Arithmetic asserts that even mixing of fecal coliforms into that assumed static amount of water would have yielded 26,000 coliforms/100 ml. That count would be about 52 times greater than the greatest count made in samples during the study.

The travel of bacteria through little grus and mostly fractured rocks can be viewed another way. Assuming half of a family's fecal coliform contribution emerged with septic-tank effluents, about 1.25×10^{11} organisms must be diluted into about 8.8×10^5 ft^3 of water to yield 500/100 ml, the highest number counted. Conservatively assuming ground waters were entirely contained within grus with 5% porosity, a land block of about 1.8×10^7 ft^3 would be needed. Because of lateral flows, it would be likely that ground waters deeper than 20 ft below the water table top would be free of added bacteria. Further, assuming the contamination formed a plume 100 ft wide, arithmetic discloses that the plume would be 1.7 mi long. There would be an additional 1.7 mi of such a coliform-laden plume each day.

Arithmetic testifies to the ability of one person's daily coliform output to contaminate such a plume of water, but accomplishing such a mixing therein obviously would be impossible. One certainly would have expected some bacterial counts higher than were obtained, but none such were had. And, as stated, while perusing Millon's data, one finds several instances of a 300 coliforms/100 ml count followed by small or zero counts 2 or 3 days later. Therefore, Millon's data testify to an extremely high disinclination of coliform bacteria to travel from disposal fields through grus or fractured granite, even a very few feet to a shallow water table, and to travel from there to elsewhere. The writer would not have hesitated to drink from the waters of wells in the Red Feather Lakes area.

Fractured Rocks and Grus: Morrison's and Allen's Studies

In 1972 Morrison and Allen [133] injected and recovered bacteria in fractured granitic and metamorphic rocks, thus documenting the capability of bacteria to be swept along with ground waters percolating through fractured rocks. Theretofore, such had been widely assumed (Figure 30).

Movement of Bacteria Through Rock Fractures Free of Ground Waters

At Cache La Poudre River Canyon, Colorado, Morrison and Allen injected tracer bacteria into highly fractured metamorphic rocks, just above a roadcut. As might be expected, bacteria were transmitted to the face of the roadcut, although not all of the bacteria safely made the trip through those open conduits.

Movement of Bacteria with Ground Waters Through Rock Fractures

At Parvin Lake, Colorado, four wells were drilled below the zone of weathering into Silver Plume Granite. It had at least three sets of joints.

Figure 30. Percolating ground waters follow fractures. Effluents percolating from a disposal field may follow fractures into a hillside to be withdrawn with well waters. Subsurface geology determines directions of ground water percolation rather than surface topography. The top end of a well higher in elevation than a disposal field may have its bottom end lower and in the percolation pathway [134,135].

The wells were located along a slightly curved line and along the expected direction of ground water flow. The distances from the first upstream well, the injection well, to Monitoring Wells #1, #2, and #3 were 19, 51, and 95 ft, respectively.

Over a 36-hr period, 55 gal of water containing 1–2 × 10^3/ml tracer bacteria were injected by a siphon into ground waters within the injection well. Bypassing Wells #2 and #3, the tracer bacteria were recovered from Monitoring Well #4, the furthest downstream. The recovery peak was 140 cells/ml after 24 hr from the onset of injection.

Assuming ground water velocities at least equal to bacterial movement, 94 ft in 24 hr is a minimum flow velocity of 0.033 cm/sec. A map indicated that the hydraulic gradient between injection and recovery wells was 0.058. Thus, Darcy's coefficient of permeability would have been about: k = 0.57 cm/sec. Experts in movement of water through porous media have classified such a permeability as borderline between "Clean gravel," and "Clean sands, clean sand and gravel mixtures" (see Table III, Chapter 3, page 29). Openwork gravels, those with unfilled voids, are extremely permeable. It would appear that the comparably permeable rock fractures amounted to open conduits.

Because injected bacteria were not recovered from Monitoring Wells #2 and #3, between the injection well and the recovery well, it could be assumed that ground water flows were highly directional. Wells #2 and #3 were only 8 and 16 ft out from a straight line connecting the injection well and Monitoring Well #3.

About two-thirds of the tracer bacteria had been siphoned during the 24-hr period from start to recovery. Thus, 37 gal containing $1–2 \times 10^3$ cells/ml had a chance to mix with ground waters. Proportions of bacteria injected and recovered lead to an estimated even dilution into from 35 to 71 ft^3 of water. Assuming the dilution water existed as a cylindrical plume between the injection well and Monitoring Well #3, that plume could have only been from 0.37 to 0.75 ft^2 in cross-sectional area. Assuming as much as 1% porosity, the plume would have occupied 37 to 75 ft^2 of cross-sectional area between injection and recovery wells in the fractured rock. Essentially, the bacteria-laden waters were transmitted in what amounted to a natural system of underground pipes.

Movement of Bacteria Through Grus

Within 2 ft of Monitoring Well #2 at Parvin Lake, a hole 2 ft deep was dug into grus. Between 20 to 25 gal of water with fluorescein dye and tracer bacteria were placed into the bottom of that shallow hole. About 10–15 min later, the water was seen to be emerging within Well #2 and trickling down its sidewall. Analyses confirmed presence of the tracer bacteria.

Within 4 and 6 ft of Monitoring Well #1, two other shallow holes were dug. Water containing dye and tracer bacteria were added to the hole 4 ft away. Water trickling into Well #1 required less than 2 hr to make the trip from that closer hole. Analyses of tracer bacteria were not presented and estimates of removals cannot be made. Nothing was said of the outcome of the hole 6 ft away from Well #1.

The velocities of flows through the grus near Wells #2 and #1, respectively, were 0.081 and 0.017 cm/sec, assuming approximately lateral flow. Actually, flow was at some downsloping pathway which would have amounted to a further distance of travel and consequently at a faster velocity than estimated. It would be reasonable to assume the waters percolated along, say, a 15° slope, or the hydraulic gradient was 0.26. Darcy's coefficient of permeability in respective Wells #2 and #1, would be 0.31 and 0.065 cm/sec. Those permeabilities are associated with, "Clean sands, clean sand and gravel mixtures," (Table III). For a frame of reference to readers more familiar with USPHS percolation tests, those respective situations were comparable to percolation test rates of about 545 in./hr (0.1 min/in.) and 200 in./hr (0.3 min/in.).

Water placed in a hole in ordinary soils will not seep out of the sidewalls of nearby holes because of surface tensions and tensile forces. It is much the same as wetting a rectangular desk blotter at the top side. Water will drip from the bottom of a saturated blotter, but it will not drip or pour out of the sides on the way down.

The writer's experience with percolation tests in grus elsewhere, together with his recollection of his inspection of the Parvin Lake site [131], leads him to believe that percolation through the grus at the study site was along preferred planes, ancient fractures, rather than through the mass as a whole.

Ancient Fractures in Grus

According to Waltz [131]—a key figure in the Colorado investigations—ground waters percolating preferentially through ancient fractures in grus can sweep bacteria along with them. (See Figures 31 and 32.) The data collected in studies of grus at the Colorado study sites, nevertheless, would indicate that bacterial travels are attended by considerable removals.

Weathering is an ongoing process. The point at which an ancient fracture could be regarded lost within a matrix of rock decomposed fully to soil is not a sharply defined event.

Solution Cavities—USGS Studies

During 1971–1974, the U.S. Geological Survey (USGS) investigated the chemical, physical, bacteriological, and virological characteristics of ground waters in five areas serviced by septic-tank systems in Dade County, Florida [122]. Hialeah was a good example of a large degree of urbanization with septic-tank systems at that time.

By 1970, nearly 175,000 septic-tank systems were discharging about 40

Figure 31. A broken rock displaying a light, fresh surface and a discolored, dark surface which had been part of a fracture plane. That fracture plane was ochre and black, colors of oxides of iron and manganese precipitated from percolation streams.

million gallons per day (mgd) of wastewaters into the Biscayne aquifer, the prime source of fresh water for Dade County. Nearly everywhere in Dade County, a solution-cavity-riddled oolitic limestone formed the upper part of the Biscayne aquifer. Disposal fields were contained within the oolite. What were termed soils were at places 48 in. thick, but soils less than 1 ft thick were commonplace.

The ability of the aquifer to transmit water was among the highest recorded in the United States. The average hydraulic conductivity ranged from 6700 to 9400 ft/day. Essentially disposal fields had been installed in a kind of system of interconnecting underground pipes containing very shallow ground waters.

Studies disclosed that as contaminant-laden waters infiltrate the ground water system, the concentration distribution is affected by dilution, chemical precipitation, filtration, adsorption, absorption, microbial assimilation, and bacterial dieoff. Advective dispersion and diffusion obscured all but the most significant additions of contaminants.

At each of five sites, a primary group of wells was dug to different depths and near the ends of disposal trenches serving one of the houses on city-sized blocks. A secondary set of wells was placed downstream of the ground water flow, 35–95 ft away from each of the five different disposal fields. In addition to each subject disposal field, all neighboring houses on the block had septic-tank systems too.

Figure 32. A roadcut through grus containing an ancient fracture. Iron ochres and sometimes black precipitates of manganese mark ancient fracture planes of original rock. Ancient fractures persist in constituting preferred pathways of ground water percolation. Sometimes plant roots grow along them.

The study was thorough. Altogether, 324 ground-water samples were collected from 42 wells ranging from 10 to 60 ft deep at the five residential sites served by septic-tank systems. Only one shallow primary monitoring well in Hialeah showed consistent bacterial and phosphorus enrichment attributable to septic-tank effluents. As was found in the studies in fractured rock in Colorado, transport in Florida's ground waters was lateral. The ground waters at the Hialeah site were about 4 ft below ground surface and flowed 3.3 ft/day. That was extremely rapid movement considering that the hydraulic gradient was only 2–3 in./mi. That shallow secondary well, 95 ft downstream of that site, was immediately adjacent to the paved street. It apparently reflected the influence of infiltrating storm waters more strongly than that of septic-tank effluents. Viruses were found in none of the monitoring wells.

It was further found [122], "Contamination from agricultural activities in Homestead and from storm-water infiltrate at West Miami, Hialeah, and north Dade are more readily identifiable and pervasive than that from septic-tank effluents."

Those USGS investigations were far more extensive than just the studies of the five sites for monitoring the immediate vicinities of septic-tank systems. Altogether nearly 19,000 chemical, physical, and bacteriological analyses were made. Viruses were studied. Ground water conditions of Dade County were delineated.

After data were in, the investigators seemed most unimpressed with what influences cultural activities had on ground waters. They noted [122], "The major influence on overall ground water quality is the nature of the aquifer."

Other Limestone Cavities

Florida's oolite is a rather recent geological formation. Oolite is still being formed by ocean chemistry in neighboring islands. Older deposits of limestone, more hardened and containing comparatively much larger solution cavities (caves), occur extensively in other Southeastern states. One might expect effective movement of bacteria in ground waters within rather large open underground holes.

A visitor noted miles of roadcuts displaying red soils only about 2–4 ft thick over thoroughly fractured, clean blocks of limestone. He inquired about well waters. That southern state's Director of Environmental Health remarked, "We have no uncontaminated wells in this state." Still, that state had no more problems with water-transmitted diseases than other states.

A sanitarian in a neighboring state noted that toilet paper had been pumped out of ground waters in limestone bedrocks where the nearest conceivable source was 20 mi away.

It would seem that Cooper was well justified in suggesting that public health authorities would better abandon their aim of zero health hazards in favor of realistic views of health risks [136,137].

A FALLACIOUS GUIDELINE: TOO FAST A PERCOLATION RATE

In various places, authorities have become concerned with local geological strata, fractured, creviced, or with solution cavities. They reasoned those openings would accept water rapidly, therefore a limit on fast percolation test rates might identify unsafe geological situations.

Soils, of course, rather than geological situations, are settings for percolation tests. Guidelines failing to deal with objectives directly find fallacious applications.

For one example rebutting belief that a fast percolation rate indicates an unsuitable soil, consider Santee, California. There a 10- to 12-ft-thick aquifer was underlaid by a much less permeable clay stratum. That ancient river bed was used to reclaim 1 mgd of secondary sewage effluents by spreading them in basins on one side of the aquifer. Percolates were recovered in an interceptor trench on the other side. Those reclaimed waters were used for recreational lakes. So the ability of the aquifer to remove viruses and other pollutants was thoroughly documented in classical demonstration studies [138,139].

The Santee aquifer was stony soil. Mechanical analyses of soil texture described sandy loams and loamy sands as having 76–85% sands, 9–18% silts, and 6% clays [140]. According to Romero's graph (Figure 28) the Santee soils would be off-scale on the safe side. Indeed, they were safe.

Lateral velocities of percolating wastewaters were about 100 ft/day. USPHS percolation tests at three locations yielded results of 23, 580, and 1900 in./hr (2.6, 0.10, and 0.032 min/in.). The geometric mean rate was 290 in./hr (0.21 min/in.). Most guidelines specifying too fast a percolation rate would have prohibited the use of the Santee aquifer for a household disposal field—that same aquifer which was proven safe even for acceptance of municipal wastewater volumes.

Waltz

Authorities in Colorado decided 5 min/in. would be a percolation rate below which certain unsafe geological formations would be found. Waltz, a geologist, was aware that the ability of a porous medium to filter bacteria and viruses was not simply a matter of percolation rate. It was an inherent quality of the medium itself [131].

Called as an expert witness in litigations, Waltz prepared two cylinders of clear plastic. Into one he placed cork stoppers which had been cut lengthwise and then slightly misfitted back together. That cylinder of cut corks, one above the other, slowly passed water containing a suspension of particles of lime. The lime was not removed and the milky water passed through.

Into the companion cylinder, Waltz placed sand. Water containing the suspension of lime particles passed through that cylinder of sand more rapidly than it did through the cylinder of corks. But, the lime was removed by the sand and the cylinder produced a clear percolate.

Waltz asserted that the ability of a porous medium to filter a percolating water is not a matter of percolation rate.

Romero

Romero also noted that the ability of a particular aquifer to remove pathogens from percolating wastewaters was an aquifer characteristic which he termed "filterability." Note Figures 27 and 28. All of Romero's collected case histories had occurred in coarse media, few of which would scarcely be regarded as soils.

Again, one could well wonder why authorities would use a percolation test in soils to identify essentially nonsoil situations. Moreover, many of the earth's most filtering-capable soils have very fast percolation rates, and some of the poorest filtering-capable soils have slow percolation rates.

U.S. EPA

In 1978, a conference was held in EPA offices in Cincinnati. Among other things, staff and consultants discussed the propriety of recommending 5 min/in. as a lowermost limit of soil acceptability for fast percolation rates. One consultant advised that such a limit had no technical merit, and it would likely rule out disposal fields in Florida, for one place, where abundant sands had faster percolation rates. Many ordinary loams also had faster rates. One of the less experienced conferees suggested 1 min/in. "How about that?" Motion carried! Scientific fact is determined neither by popular vote nor authoritative decree, but some guidelines are.

In 1980 the EPA's *Design Manual Onsite Wastewater Treatment and Disposal Systems* [16] noted, "Soils with percolation rate $<$ 1 min/in. can be used if the soil is replaced with a suitably thick ($>$2 ft) layer of loamy sand or sand," (which incidentally, can easily have a percolation test rate faster than 1 min/in.). One can well imagine digging a disposal trench in the grus (itself at least as coarse as other sands) at Parvin Lake. The trench could be 4 ft deep, contain the EPA recommended sand to 2 ft deep, then have 1 ft of gravel and 1 ft of earth backfill. Effluent could flow into the trench, out the sidewalls, and follow the fractures to Kingdom come.

Recommendations

Do not use a percolation test to identify a geological setting. There is no substitute for proper site investigations by trained technicians. Rulings on too fast a percolation rate are arbitrary, capricious, and take the place of proper studies.

MINERALS FROM SEPTIC-TANK SYSTEMS

Nitrates

The environmental movement saw a confusion of value systems with concerns of public health. Notably, an extremely rare, largely outdated, and easily remedied malady—methemoglobinemia (commonly the blue baby disease)—became justification for a nitrate standard. The political potency of that tool has been obvious. Not so obvious has been the technically weak justification for its being. The reader is referred to a technical treatise on the topic [141]. He is also informed of Wert's measurements of nitrogen removal in a disposal trench system [142], available since the writing of the referred treatise. Wert's measurements add to other measurements contradicting common belief that nitrates from disposal fields travel extensively through soils.

Phosphates

Chen [5] has discussed the chemistry of phosphates in soil systems. He presented six known chemical reactions producing relatively insoluble precipitates of phosphate in soils.

Sawhney and Starr [143] studied movement of phosphorus from a septic-tank disposal field. They found only 0.5 mg/l of phophorus (98% removal) about 2 ft under a 6-year-old disposal field. They also found, ". . . resting of the system regenerates phosphorus sorption sites in the soil and then the soil is capable of removing additional phosphorus from wastewater over a longer period of time." That constitutes another benefit from alternation of fields.

Sawhney [144] reported, that the soil in which the subject septic-tank system had been situated was a fine sandy loam with fewer than 10% clay particles. Such soils tend to have lower sorption capabilities, which in turn would be expected to provide fewer cations for phosphate removal through precipitation. Still, removal had been 98% within only 2 ft of the bottom of the subject disposal field. Removal was likely complete after effluents percolated only a few feet farther.

It is general knowledge among farmers: it is a problem to get phosphate in fertilizers to move to the root zone.

Heavy Metals

Heavy metals are not a normal constituent in household wastes. Still, sometimes a household hobby will yield some heavy metals to a septic-tank system.

When most metals are in soluble form they are electrically attracted to soil particles and become a part of the ion-exchange complex. Arsenic is an exception, and finding it in ground waters of desert areas is disquietingly frequent. Well waters in such areas should be and generally are analyzed for mineral contents.

To a plant physiologist, it seems unrealistic to expect plant roots failing to take up most minerals from percolating household wastewaters. On a hillside where percolation is largely downslope in a thin topsoil, plants have every opportunity to utilize minerals passing therethrough. Even when heavy metals are precipitated or a part of the soil ion-exchange complex and not free to travel with percolating waters, plant roots will take them up. Plants take up silicon from sand. They even accumulate gold when it is in soils.

In passing, it should be pointed out that plant roots do not act simply as wicks, taking up waters and dissolved substances indiscriminately. Mineral uptake is an active, metabolically controlled process. Substances coarser than dissolved minerals, e.g., bacteria and viruses, should not be expected to enter an intact plant root along with water uptake. Microbes can at times enter broken roots. Sometimes, bacteria and viruses can actively attack a plant, host to their specific pathogenic activities. Ordinarily, however, microbes pathogenic to mankind are not to be found in plants growing over disposal fields.

Boron

At high boron concentrations, plants can be killed. At lesser concentrations, symptoms of toxicity other than death may be seen. At still lower concentrations, optimum growth is had. Below that optimum boron level, deficiency symptoms occur. As would be expected, plants vary considerably in their boron needs and tolerances [145]. As opposed to plants, most of animal life is relatively tolerant of boron.

In the *1955 Yearbook of Agriculture* [146] the USDA advised, "Toxicity may develop with boron-sensitive crops when the concentration is as low as one part per million (1 ppm). Water containing 1 ppm or less of boron may be regarded as excellent. For most crops, however, water that contains 1 to 2 ppm is satisfactory. Water up to 3 ppm may be used with the more boron-tolerant crops. Water containing more than 3 ppm is doubtful or definitely unsuitable for irrigation purposes."

Some products for household washing use contain boron. But ordinarily not everyone in a community will use the same product for household washing needs. Consequently, collected community wastewaters generally have concentrations of boron below levels deleterious to most plants.

But when a household has a septic-tank system, daily wastewater discharges will completely reflect household choices of products which are ultimately influent to the septic tank.

Products containing boron for laundry needs used as directed could result in concentrations in a septic-tank effluent in excess of levels of tolerances of many plants. Concentrations could vary from a low of 4 mg/l to highs of 34 mg/l boron or more. Influent concentrations would be much higher, but a septic tank acts to buffer incoming flows. Effluents are far more even in qualities than influents.

Septic-tank systems serving a fourth of our nation [147] are installed in about every conceivable situation. In some places, percolates from disposal fields would rapidly leave root zones of plants. In other situations, however, plant roots would have every chance to be exposed to effluent percolation. A disposal pit in desert soils would be a good example of the former situation. The latter situation would be had in sloping, hilly terrain where percolation was downslope in a permeable topsoil overlaying a less permeable substratum.

These remarks stem from knowledge of plant physiology and septic-tank technology. It must be pointed out, nevertheless, that no pertinent field studies are known to the writer demonstrating a situation deleterious to plants existing in the field.

Nonwastewater-Oriented Mineral Pollution of Surface Waters Which Could Recharge an Aquifer

In 1974, melting snows in the San Bernardino mountains flowed into Lytle Creek, California. The snows must have been at about 10,000 ft of elevation. Creek waters were sampled at about 3600 ft, before cultural influences could affect the running snowmelt. About 2 mg/l of nitrates were contained in the snowmelt which arithmetic predicted would be expected of precipitation scrubbing actions through smog [148].

COMMENTS

At the onset of this Chapter, the inclination of people to ponder the trivial while ignoring matters of moment was pointed out. In Santa Clara County, California, where one may hunt ducks in sewages, a sanitarian stood on a mountainside. At his feet was the subdivider's effort to prove the land feasible for disposal fields. A trench had been dug with a Caterpillar D-9 with a smoking ripper tooth into somewhat fractured but unweathered serpentine. That situation was of no concern. The sanitarian looked a good portion of a mile into the valley below. He pointed to a creek which had been posted for its dangerous and always-present mercury contents. He pondered, "What about the nitrates from the septic tanks reaching that creek and polluting it?"

The obvious is elusive when it is constantly present and lacks the surges needed to cross human threshholds of genuine consciousness. Thus, persons in government encouraging endless accumulations of people in ever-growing cities fail to see that their plans would be about the same as if they were deliberately trying to increase crime rates and other high-density, high-population associated maladies. Authorities of public health know that major epidemics have come from water supplies, sewers, and other appurtenances of population centers. Yet those same health authorities often encourage sewering and other known density- and growth-inducing public works [149]. Even primitive peoples knew to separate when epidemics came.

Thanks to good public sanitation, good food, good housing, and other aspects of prosperity, our citizens enjoy generally good physical health (mental health notwithstanding). Prickett and Cooper [150] estimated that statistics of waterborne infectious hepatitis predict 90 persons to be infected each year in the United States. At the same time there were 19.5 millions of septic-tank systems, cesspools, pit privies, and "other" on-site means of sewage management in the United States in 1970.* To an objective observer from outer space, it would appear obvious that if on-site sewage management constituted health hazards at all in our country, they would be monumentally insignificant.

---

* According to the United States Bureau of the Census, in 1970, 71.2% of our All-Year-Round Housing Units were served by a Public Sewer, 24.5% were served by Septic Tank and Cesspool, and 4.3% were served by "Other" [147].

CHAPTER 11

SITE EVALUATION

Few sewage disposal problems are technical. Most are people problems. In site evaluations, the current state of human affairs, their codes, and predispositions should be assessed first. Technical considerations are subordinate, and a feasibility study of a problem must be aimed at politically permissible solutions.

POLITICAL CLIMATE AND OTHER HUMAN FRAILTIES

Above all, the most critical site feature is the political climate—the relationships between the public, the local authorities, the client, and the consultant. Most technical aspects of waste disposal problems are simple, but it is rare that problems can be solved by technical considerations alone. Local codes, authoritative dispositions, and other man-provided aspects of problems come first. Thus it is pointless to consider water savings in wastewater estimates where local authorities mandate design estimates independent of actual expectations. And, it is pointless to explore soils in deep holes if disposal fields are mandated to be shallow, and so on.

Value systems, always believed to be universally held by each of us, differ. Just a few years ago during the height of the environmental movement, there were two proposed campgrounds in Southern California. Both were in high mountain meadows. Both were essentially alike, but were in different counties. In one, the local concern was for the trees (later killed by construction, to the concern of no one). There, consultants were obliged to place campsites beyond the trees in the open meadow. In the other county, the local concern was for the meadow. There, consultants were obliged to place campsites under the trees and out of the meadow. Social schizophrenia very often determines where a consultant is to locate facilities, and his soils investigations must be so guided.

The human tendency to think along well trodden pathways is universal. In a state park in the Pacific Northwest, a disposal bed had been a failure and no efforts to repair much helped. Soils were very slowly permeable. A consultant from out of state was called. Upon being driven to the park at the top of a hill, the consultant looked at the roadcuts along the way out of the automobile window while his hosts discussed a history of problems with the site. At the top of the hill, a backhoe waited. When everybody arrived, the backhoe began digging a hole in a candidate area for a repair system. When the hole was as deep as customary disposal fields were, the locals had seen enough of practically impermeable soils, and they were ready to dig elsewhere. The consultant said, "Keep digging." Astonished at his wish, the backhoe dug a few feet deeper. Soon the hole displayed a deep stratum of very permeable sandy soil. It had been displayed in the roadcuts, but only the outsider looked.

Human frailties are always a critical part of site evaluations. In a subdivision situation, an outside consultant was called to solve what was believed to be a percolation test problem in obviously suitable soils. The subdivision engineer had misspelled the name of the local authority. That was the real problem, but the local authority would not openly admit to it.

A project's financial strength is another aspect of site evaluation. If strong enough, the consultant may entice authorities to allow him to try new tools, provided: (1) he documents their feasibility, (2) the project guarantees the constant surveillance of construction of the system by the consultant, and (3) ongoing responsible involvement of the consultant is assured to the time of project independence.

It is difficult to convey the paramount importance of human nature as a site factor to be considered, and efforts to do so terminate with this last example.

A small mountain community faced local county government with hostility. The county effort was to solve sewage disposal problems. It appeared that the locals wanted only to be left to their troubles. Nevertheless, an outside consultant became accepted by the community, his being regarded as separate of county government. The problem? Quite simply, county staff had had several changes and the local people were out of patience with seeing new faces, new going-to-solve-your-problems attitudes. Nothing more than failure to realize the need for human continuity caused political difficulties which have blocked proper actions on problems to date.

LOT SIZE AND ASPECT

Political

Especially in recent years, our government prefers to plan our destiny rather than to allow it to happen organically. Cities like San Francisco, Santa Fe, New Orleans—cities people travel from all over the world to see—can no longer be built. They violate almost all rules, regulations, and other of today's governmental controls. We are now limited by the imagination of our planners, and we can look forward to relief from aesthetically boring communities only by the variances someone gets away with. And along with other aspects of government-mandated environments, there is the question of lot size.

Rather than face the obviously value system-oriented wish to control lot size, many governmental authorities foist the problem of controlling lot sizes onto the local health department, provided individual wastewater disposal is an available justification. One state health department was compelled to find reasons a lot size was to be larger than use of septic-tank systems could justify. It was decided to believe that all nitrogen disposed to septic tanks would reach deep ground water basins. Calculations based on an estimated ground water volume and an assumed human tolerance limit to nitrates were then used to estimate the number of septic-tank systems each area could contain. Lot size was so justified. But, of course, the situation is subject to obvious attack [141]. It only remains for a politically strong enough opponent to knock the technically flimsy lot-size justification down, and some credibility of the health department with it.

Technical

Technically, a lot needs to be only large enough to contain the septic-tank system and man's other improvements. Despite slowly permeable soils, the largest of mandated septic-tank systems to be placed on an authoritatively acceptable soil is not but a fraction of an acre in size. But, a designer cannot look at lot size from that point of view alone.

In more affluent areas, a house with its garage, pool, tennis court, and the like can easily crowd an acre or so. An acre of open land looks rather large to a consultant performing percolation tests on a pleasant day, but a return visit during house construction will always be a shock. He will well wonder if the disposal field, will fit in at all.

In hilly terrain, the designer must account for a driveway and other excavations. Where permeable soils and less permeable substrata are intersected by a cut, a disposal field located too closely above may well provide effluents percolating to daylight in the cut below. The lot must be large enough for avoidance of such a sitution, or perhaps the designer can demand plan changes. Sometimes, a cut can be avoided by stipulations to a driveway on fill only.

Steep terrain, setbacks, easements, aesthetically valuable trees, and other considerations can render space on a lot valueless to a designer of a septic-tank system.

Psychological

Generally, as will be discussed in Chapter 12, where lot sizes are large enough to separate neighbors from their neighbors' problems, lot size is psychologically large enough. It would be a mistake to believe that surfaced septic-tank effluents are anything less than one of the most severe nuisances known to mankind, not so much our own effluents—someone else's. A designer should look to slopes, to effluent-hiding capability of dense undergrowth downslope of the contemplated disposal field, and other aspects of the lot which conceivably could bring to or hide effluents from a neighbor's attention. All disposal fields should be planned for best chances of successful operations forever, because such unrealistic perfection is demanded. Nevertheless, the designer would well remember that he has no control over the water-wasting habits of a household and other unpredictable matters of the future. He would best look at the planned disposal area and imagine the outcome if problems did arise one day. Nuisance to neighbors is the thought, because by realistic definition: septic-tank failure = a telephone call of complaint to the local health department.

SEPTIC-TANK PRACTICES VS PUBLIC SEWERS

When a private consultant views a site for a contemplated subdivision, before considering septic-tank practices he should consider the site's proximity to the nearest sanitation district, whether or not the subdivision would justify an in-tract sewer and disposal facility, and other aspects of deeply rooted attitudes favoring sewers.

An expected rate of development has engineering significance, but local authorities have a tendency to view a project as ultimately developed. The number of lots, their size, likely affluence of the community yet to be, all should be viewed by the consultant as though through the eyes of a person with emotional inclinations towards sewers.

After all, authorities to grant approvals for subdivisions with septic-tank practices are directly responsible for septic-tank practices thereafter. Public sewers, on the other hand, fall within the purview of other entities which benefit by those devices.

It should be borne in mind that the presence of septic-tank systems or sewers is practically soils independent. No one needs to inquire if Los Angeles is on sewers, sands below notwithstanding, or if a lonely house with pressurized water is on a septic tank (or nothing), its being on a rock notwithstanding. The inclination towards sewers is strictly a matter of politics and economics. Health hazards, environmental concerns, community growth, all are topics of concern, but none are really decision-making factors.

HISTORY OF EXISTING SEPTIC-TANK SYSTEMS

Sometimes a subject site has neighbors with existing septic-tank systems in same soils and terrain. The history of performance of those systems can offer valuable information.

When the project is large enough to justify the efforts, a consultant can canvass the neighbors. With patience and skills discussed in Chapter 12, the troubles, or lack of same, with systems can be documented. Oftentimes samples of soils near the existing disposal fields can be collected and subjected to mechanical anaylsis of texture [37,151]. Those analyses can be compared to analyses of soils sampled at the subject site to display the degree to which they are alike. It may be argued that successful septic-tank systems on nearby same soils would predict successful practices in the subject project. If a USDA Soil Conservation Service office is nearby, a soils scientist may volunteer his aid in relating the neighbor's soils to those of the subject site. A soils map may also help, but it should be understood that those maps can be in error, particularly in the western states where huge land areas may have only slight agricultural value and where soils vary much in small areas.

When canvassing neighbors, history of their systems can offer useful design information. At one subdivision, for example, many neighboring septic-tank systems in same soils had served trouble-free for many years beyond expectations. But, the terrain and soils situations failed to conform with new regional guidelines. Proof, plus the political voice of a fair-minded representative, enticed regional authorities to accept physical reality. Guidelines were varied in the proposed subdivision to accommodate septic-tank practices there. The consultant had found a few troublesome septic-tank systems belonging to neighbors. Without exception, those systems were in the almost flat terrain of natural subsurface drainage swales. That finding led the consultant to avoid such swales or to recommend curtain drains where they could rectify matters.

SOILS

As interesting as soils are and as helpful as soil sciences can be, there are only four paramount soil considerations pertinent to septic-tank practices: (1) soil permeability, (2) liability to soil damage, (3) soil extent, and (4) topographical soil aspect.

Soil Terminology

Terminology of soil science has found its way into septic-tank practices. Several local authorities unaware of the special skills needed in performing percolation tests, unaware of the enormous damages disposal fields suffer during construction in wet weather, unable to obtain reliable information from industry,* or otherwise unable to account for problems, have looked to soil scientists for help. In other areas, geologists have accepted the same rescue role. At the same time, soil scientists have grown tired of mapping soils for the USDA Soil Conservation Service, and they have been waiting for a chance to aim their talents towards the greater population—people not in agriculture. Consequently, local authorities are rolling clods of soil in their hands, making ribbons of soil, biting it between their teeth to judge sand content, and much enjoying a very interesting discipline.

A disposal field depends on a porous medium for function. Whether or not the medium is termed soil, subsoil, weathered rock, or highly fractured rock is of interest to soil scientists and geologists. Such terminology is not of much concern to the designer of a septic-tank system. It is only important to know that a disposal field may be expected to inject effluents into a porous medium and the medium can assimilate those effluents in accordance with proper considerations of public health.

Still, people do have a tendency to define an item. Then they get lost in their own semantics while overlooking the item's function. In California, where strange things can happen, one state authority even provided a

* Many health departments accept percolation test results from contractors who may or may not have a job, depending on test results. One contractor put it quite well to investigators [21], "We give them whatever results they want." Not all contractors fabricate acceptable percolation test results. Some are willing to lose business, but such unfair situations of conflicts of interest should be avoided.

laboratory analysis for distinguishing soil from rock. One technician collected a sample of somewhat porous bedrock and found it to qualify as soil.

In accordance with soil scientists, *soil* herein refers to what will pass through a 2-mm sieve (between 1/16 and 1/8 in.) [37]. That definition best serves the designer of a disposal field, because it is the finer particles rather than the rocks which are associated with removal of bacteria and viruses, soil structure on which permeability depends, and the like.

Soil texture refers to the proportions of the various particles of which a soil is made—sand, silt, and clay being the coarsest to the finest particles. (See Figure 33.) The various assortments of particles give rise to the expressions of textural classes—loam, clay, sandy loam, etc.

Engineers and soil scientists have slightly different textural classifications. Soil scientists have classified sands being from 2 mm to 50 μ, silts being from 50 to 2 μ, and clays being smaller than 2 μ [37].

Textures can be subjectively judged by experienced persons, and objective determinations can be made in a laboratory by mechanical analysis of texture. Sieves are used for coarse particles, but a hydrometer technique is customary for determining quantities of fine particles.

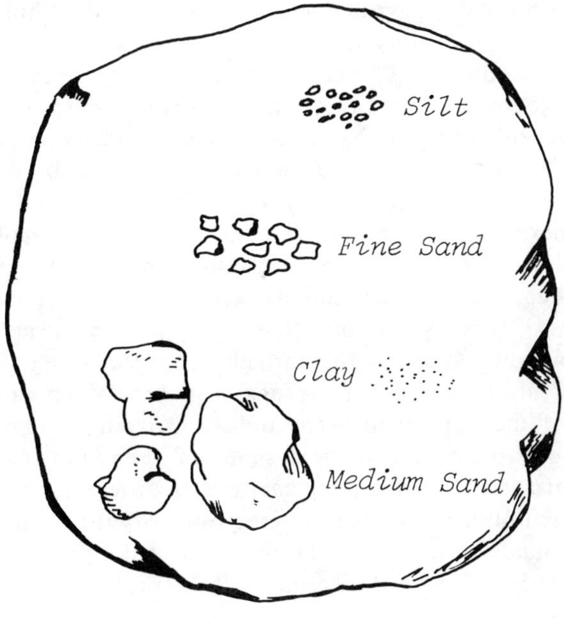

Figure 33. Sand, silt, and clay depicted upon a grain of coarse sand.

Engineering laboratories will provide a graph with proportions of all particles they receive. But samples with sizeable rocks are not representative. It is best to instruct the laboratory to analyze only the 2-mm fraction, that is, the portion passing a No. 10 sieve.

With the exception of clean sands, soils are not simply piles of particles. Those particles are arranged by virtue of the ability of the finer particles and organic matter to stick together. The arrangement of particles is termed *soil structure.*

Structure is critically important to septic-tank practices. It controls permeability, and in that respect it constitutes the liability of a soil to damage. In Chapter 3, it was maintenance of soil structure which was in mind when discussing care in digging and water-filling percolation test holes.

"*Soil drainage*, as a condition of the soil, refers to the frequency and duration of periods when the soil is free of saturation or partial saturation. Soil drainage, in a dynamic or active sense, refers to the rapidity and extent of the removal of water from the soil, in relation to additions, especially by surface runoff and by flow through the soil to underground spaces," [37]. With respect to soil drainage, the reddish colors of some soils (ochres of oxidized iron), prevalent in the South, result from good soil drainage and aeration. Septic-tank technicians have long noted, "The redder the better."

Aspect of soil as discussed herein refers to the lay of the land, its slope, its vegetation, its management of ground waters by virtue of aspect, and its underlying geology insofar as it could be considered a part of the greater porous medium into which wastewaters are to be injected.

Permeability refers to the ability of a soil to transmit water as measured under strictly controlled circumstances. The percolation test does not measure permeability; rather, that empirical test yields measurements which, among other factors, are a result of permeability.

Soil science is interesting; but that discipline has been aimed at agricultural pursuits. Whereas the agriculturalist can settle quite nicely with enough soil permeability for crops, a disposal field functions on about 2–3% of the initial soil permeability. Clogging is expected. So, when speaking of permeability, the designer of a septic-tank system and the soil scientist are understanding matters of different magnitudes.

These few definitions have been offered to clarify their use herein. No effort will be made to introduce the reader to soil sciences, that being beyond the scope and needs of this volume. Still, the reader is encouraged to learn more about the interesting subject of soil sciences. He is encouraged to visit his local USDA Soil Conservation Service office and to request guidance to a better understanding of soils from the local soil scientist. Those gentlemen are invariably pleased to help. Additionally, the EPA *Design Manual* [16] has an excellent presentation on soils with photographs and drawings. (This recommendation is not to be construed as a recommendation for the remaining sections of that treatise.)

Fallacious Use of Soil Maps

Soil maps in our country have been historically prepared for agricultural purposes. Accuracies are proportionate to need. In Eastern areas where small farms are found, more accuracy has been needed than on a Western ranch with many square miles of almost worthless rangeland. For use in septic-tank practices, soil maps should not be accepted at face value. Years of analyzing soil samples from Western subdivisions have caused the writer to ignore soil maps entirely. They have too consistently lacked needed accuracy. Permeability and extent of a soil are the concerns and such information best comes from site explorations.

The ratings offered by the USDA Soil Conservation Service as to feasibility of septic-tank practices should be ignored. Not only do disposal fields serve quite nicely in soils noted "severe" in limitation, but the concept of using soils maps to predict the outcome of septic-tank practices is fallacious. The soil is a natural building material. Human structures will be good or bad, depending on how well the designer accounts for the building materials he uses.

The soils map approach to septic-tank practices was an FHA-supported research project at Virginia Polytechnic Institute. The investigators, R. D. Krebs and J. H. Hunter, evaluated septic-tank practices and compared them to various features of soils [152,153]. The hope was that soil maps could be used to predict outcome of practices.

It was found that very good and very poor soils situations could be identified, but the remaining soils remained elusive and conditional. It became obvious that in the conditional soils, design, construction, inspection, and maintenance of septic-tank systems governed outcomes. For example, good histories of septic-tank practices were had in a certain soil in a county. Poor histories were had in the adjacent county. The soil did not change at the county line, but septic-tank practices did.

Despite the negative findings of those FHA-sponsored researches, soils maps have been used to guide septic-tank practices in some places. Once when visiting a county health department in the Midwest, a homebuilder was seen to come in for a septic-tank permit. The sanitarian looked at a soil map, named the soil and its texture, sized the contemplated disposal field, issued the permit, and sent the applicant on his way. The eavesdropping visitor inquired about that approach to design. The sanitarian assured that the map had been specially prepared (at goodly cost) for the health department for that use, and if any question arises, the sanitarian will visit the site and check the soil classification himself. It happened that the eavesdropping visitor visited the particular site later, and that trained visitor did not find the soil as classified. Further, its aspect was such that troubles with an ordinary design would be predicted.

One may well wonder why the soils map approach works for his situation. It happens that septic-tank systems are far more tolerant of abuse than realized. The fact is that whether or not a soils map, a percolation test, or a divining rod is used, septic-tank practices will be however successful they generally are in an area. The local authority has no way to know if the soils map approach works or not.

Nevertheless, the use of a soils map is administratively easy. Having decisions mapped out would be a genuine help. But use of the soils map approach is technically unsound practice.

Soil Profiles

Roadcuts often are helpful in displaying soils and rocks in profile. Nevertheless, soils slough on roadcuts and colors change as time passes. The technician can use roadcuts, but he must be wary and ready to dig for his findings.

Geological formations change little and, aside from damage from rock-breaking machinery, roadcuts provide excellent displays of geological units (see Figure 34).

A backhoe is an excellent, and this writer believes indispensable, tool for site investigations. Soil depths, soil profiles, and presence of high ground waters, all can be best studied with the aid of a backhoe. Having selected a spot, the backhoe should dig parallel to contours rather than at right angles to them. The hole should be about as deep as the machine can dig, or until substrata are so hard as to refuse further digging.

A ramp or series of steps can be dug at the end nearest the backhoe, and a man can enter the hole. With the advent of the Occupational Safety and Health Administration (OSHA) (and consequent increase in accidents now that everyone is told he's safe) a technician might need the hole wide enough at the top so that collapse could not occur. Some technicians can from experience judge the stability of the hole wall. Generally, when a soil is freshly dug, it is far more stable than a few days later. Any hole left open a few days should be regarded as dangerous.

During digging, ground waters can sometimes be seen seeping into the hole from its sidewall. Although the level of seepage might be seen and measured, the best way to ascertain ground water levels is to allow the hole to water-fill to piezometric level (the stable level of the water within the hole). In a small hand-augered hole, piezometric levels may be

Figure 34. Mother Nature is not neat. Finding these alternate strata of soils and rocks along a roadcut suggests use of deep trenches so as to intersect as many useful strata as practical.

reached rather readily, but a backhoe hole may well require overnight or longer.

The sidewalls of a backhoe hole are generally much damaged, although not always obviously so. It has been found best to cut a vertical V-shaped notch with a knife, or otherwise remove an inch or so of the sidewall surface to be viewed. One technician used a hoof-pick quite effectively, and it served more quickly than a knife for soil removal. A hoof-pick is used for working on horse hooves and is available in saddlery and harness shops, tack rooms, etc.

Tree or shrub roots can go quite deep, and their presence is an indication of permeability and aeration, depending on the plant species. Grasses and herbs are generally shallow rooted, and deep roots in grasslands should not ordinarily be expected.

When soils are waterlogged for extended periods of time, they may (or may not) change colors in a mottled fashion. A view of a soil in profile will usually display shades of the same color, although wet areas are generally darker. If mottles are present, one will see one color, usually yellow to ochre, in a matrix of another color, perhaps grey. A soil scientist acquainted with the particular area might know whether or not the mottles indicate seasonal ground waters or exist for other reasons, but investigations during the season of rains are best to determine such.

Technicians vary much in their recording (logging) of their visual examinations. Generally, it is useful to know:

1. depth to which the hole was dug before refusal, if any, was had to the backhoe
2. depth to which plant roots were found
3. depth to ground waters, if present
4. depth to mottled strata and their thickness, if present

In addition, some technicians who routinely perform percolation tests can judge approximate permeability. Soil textures of different strata might be judged and noted.

The technician must judge how many backhoe holes are needed for an adequate sampling. It generally cannot be assumed that a single hole is sufficient in an unknown situation. After all, a big hole might be, say, 3 ft wide and 12 ft long. That occupies only about 36 ft^2, or about 0.0008 acre. At the same time, however, one does not want so many holes as to damage the site for practical purposes.

During explorations, the technician should remember that soils backfilled into a test hole are not compacted to a natural degree. Houses constructed over such holes have yielded lawsuits after cracks developed. All such holes should be located on maps. Further, those holes serve quite nicely for percolation tests along their ramps. Thus they are often left open overnight. Insurance agents pay claims for people and horses, but as a rule cattle are intelligent creatures and they will not walk into a gaping hole in the ground. Since humans value animal life in terms of size, small critters that fall into holes usually excite no lawyers.

Biological Zone

Soils are not static. They are always changing. Particularly, soil microbes are constantly interacting with particles, gases, organic matter, and in about every imaginable way. Soil microbes react to wastewaters. They utilize the wastes as foodstuffs and discharge by-products into the percolate stream. The net effect is eventual wastewater renovation.

Sometimes authorities question the wisdom of constructing disposal pits deeply beyond the biologically active soil zone. Those concerns can be laid to rest. Bacteria are always associated with tree roots, and they have been found 100 ft deep in fractured rocks [154]. Iron-reducing bacteria have been found in wells 1000 ft deep [155]. Aside from microbes being in deep soils at the onset, wastewaters carry their own microbes and foodstuffs to whatever depth a disposal field happens to be.

Occasionally there is reluctance to design deep pits for concern that atmospheric oxygen will be excluded from soils of much depth and anaerobic conditions would be had. Indeed, oxygen is needed for final satisfaction of residual biological oxygen demands. The concern seems groundless because the soil atmosphere has very closely the same gaseous makeup as the overhead atmosphere. Barometric fluctuations cause atmospheric interchanges to considerable soil depths. Witness oxygen-dependent tree roots found as much as 100 ft deep in some places.

Soil Permeability—The First Paramount Consideration

During their two-decade studies of septic-tank systems, investigators of the USPHS determined performance histories of thousands of systems. Every measurable soil characteristic believed even remotely related to performance was studied. Always, only soil permeability much mattered, and that in terms of a percolation test.

It may be of personal interst to observe soil texture, to identify clay loams, silt loams, and other textural classifications. But texture was only one of the soil characteristics weakly related to performance of disposal fields; that is, texture was unreliably related to permeability. Percolation tests should be performed. Visual judgments, nevertheless, or the miniperc (Chapter 3) can offer valuable planning information prior to a percolation test program.

When considerations of assimilation depend on estimates of soil permeability, the reader has been referred to a simple falling-head permeameter [87]. Where convenient or representative soil samples cannot be collected for permeameter measurements, the reader has been advised to use a formula of correlation which relates the USPHS percolation test performed in a certain way to permeameter measurements. (See Chapter 3, page 30.)

Liability to Soil Damage—the Second Paramount Consideration

There are two kinds of soil liability to damage to observe in the field—that is, liability to smearing and compacting, and liability to migration.

When the backhoe digs, teeth make grooves in the far end of the hole. The technician can use the miniperc to compare residual permeability of the backhoe's teeth marks to a fresh soil face of a broken clod. The disparity will give an indication of liability of the soil to loss of permeability because of mechanical damage. Also, the technician can note whether or not the soil is clayish and therefore has a permeability almost solely dependent on a maintained soil structure.

Except in clean sands, there will be some soil damage from digging, and the technician will need much experience to judge the severity of the liability. As a rule, clays and clay loams will suffer most damage, and moisture will contribute.

Where damage-liable soils are found suitably permeable for a disposal field and other site considerations are favorable, it is best to recommend that all construction of disposal fields be performed only during the drier season of the year. Advice can be obtained from the local USDA Soil Conservation Service if the proper, drier time of the year is not known to an anxious builder. If for some reason the recommendation is questioned or that likelihood foreseen, perform a careful percolation test. Then, nearby in the same soil, dig another hole with an auger or posthole digger. Provide no paper basket liner, and pour water fillings from a bucket above. The disparity in resulting percolation rates will be convincing.

It was mentioned earlier that some silty soils lacked cohesive strength and water-filled percolation test holes would display soil migration. And the USPHS observed silty soils having infiltrated gravel fillings of disposal fields. Sands are capable of about the same behavior. But sand grains may not be carried so far as lighter, smaller silts, and sands will seek their angle of repose in a gravel-filled disposal field [44].

Beside soil migration during percolation tests, silts can be identified by standard hydrometer techniques, mechanical analysis of texture [151]. In the field, silty soils leave hands with a feeling of having handled flour, and when dry, hands can be brushed relatively clean. Clay on hands must be removed with soap and water. Clays feel sticky when moist. A dry clod of a silty soil can generally be crushed between thumb and forefinger with modest effort. Dry clods of clay cannot. Still, texture properly identified or not, a soil which will migrate is best identified by observing overnight migration into a percolation test hole filled with presoaking waters (usually a paper basket within the hole will give way to migrating soils).

Soil migration will prematurely terminate the lifespan of the ordinary gravel-filled disposal field. Special designs will be needed (see Chapter 14).

Soil Extent and Soil Aspect— The Third and Fourth Paramount Considerations

Once septic-tank effluents have been successfully injected into a natural soils system, those effluents must be assimilated. That requires considerations of the extent of a soil deposit and its topographical aspect.

Little help is offered in most local guidelines. Probably because soil deposits are generally adequate for single-family residences and judgments of soil extent and aspect require subjective skills, guidelines are customarily limited to a request for a ground-water-free soil depth below proposed bottoms of disposal fields and indirectly pertinent requests for setback distances from wells, streams, property lines, etc.

Three conditions of soil aspect should be known:

1. depth to less permeable substrata
2. comparative permeabilities of soil and substrata
3. slopes

If soil aspect is flat and sufficiently deep above a less-permeable substratum or a ground water table, percolating effluents from a disposal field might mound adequately for needed hydraulic gradient [6]. If a poorly permeable substratum is free of a ground water table, it would appear that the permeability of that substratum might be enough to serve. And as the terrain slopes more, so also does the hydraulic gradient become more favorable. The reader interested in formal methods of estimating the net effect of these three conditions can find formal help [6], and informal help appears in Chapter 9.

Where a shallow substratum underlaying a more permeable topsoil has not enough permeability to transmit rainwaters seasonally, a seasonally perched water table may exist. Besides soil mottling, vegetation may betray such a situation. Trees may have a stunted appearance. Oftentimes cattails will be seen in flatter areas below sloping terrain. There, a quite permeable hillside soil may drain to low, almost flat spots, that drainage being adequate for disposal fields on slopes above. A less permeable hillside soil, on the other hand, may contain seasonally perched ground waters which cannot drain downslopes fast enough to prevent perching.

For estimates of assimilative capacity, measurements of permeabilities of soil and subsoil and degree of slope should be made. Potential evapotranspiration is also a site factor for the technician to evaluate. This is not to recommend that the process be relied on. It is to note that that process occurs and can be a positive contribution when disposal trenches are at the top of a slope wherein evapotranspiration may dispell downslope percolating waters.

Where permeable soils are shallow over much less permeable substrata, the construction of a roadway, even a gravelled roadway, can create in effect a compacted soil dam to subsurface ground water movement. Whereas an area may be capable of transmitting rainwaters throughout a shallow soil, construction of roads essentially can create a system of isolated soil blocks, especially in almost flat lands [156]. Where the technician finds that aspect of soils, he should inform planners of the need to arrange roadways accordingly and to somehow construct them such as not to create dams to ground water flows.

Before the advent of popularity of the concrete-slab house, houses were elevated a couple of feet or so to have wooden floor joists well above ground. Then if septic tanks had effluents surfacing, they were in the yard. The slab house, on the other hand, is already at about ground level. Plumbing is supposed to be at a given downgrade, and it frequently is just able to join a shallowly placed disposal field. Without a driving hydraulic head of a few feet, effluent surges from a septic tank need only to meet with a slight resistance from an ordinarily suitable disposal field to cause backing up the line to the house. Ladies tend to object to sewages from the toilet pouring onto their oriental carpets.

When finding shallow, less permeable subsoils in flatlands, the technician should be aware of the need for enough hydraulic gradient from house to a shallow disposal field to have a satisfactory system. Where permeable soils can accommodate a disposal field but are too shallow for proper plumbing, serving the field with a sump and pump may be advisable.*

Occasionally in desert areas, and perhaps elsewhere, a soil deposit overlays a depression or basin in an almost impermeable substratum. If that basin prevents wastewater from escaping, a disposal field can water-

* In 1977, studies of practices in Florida disclosed ordinarily functional disposal fields yielding occasional troubles because of inadequate considerations of hydraulics. National authorities providing guidelines for septic-tank practices, then and now, have not considered Florida's very different and peculiar soils and geology. Thus, following nationally provided guidelines in Florida, or in any other special geologic setting, will constitute misfitted technology [44].

fill the basin and disable the field. Such situations are not generally betrayed by a pattern of exploratory, deep, backhoe holes. The technician needs to be aware of the possibility of a subsurface basin, and geological skills may be needed if suspicions are aroused.

When viewing soils in profile, the technician should look for openwork gravels (gravel without soil-filled interstices), open fractures in rocks, and other open conduits for transmission of unfiltered septic-tank effluents to water supplies. Gravel strata and rock fractures filled with soils offer filtration.

Oftentimes, bedrocks (granites especially) have weathered such as to leave large boulders and large rocks within a matrix of soils. From the technical viewpoint, those soils are deeper than the rocks, provided percolates from a contemplated disposal field could be transmitted around those rocks. Oftentimes a backhoe can dig very deeply beside a rock outcropping, thereby demonstrating the useful depth of soils (see Figure 35).

In mountainous terrain, roadcuts and backhoe holes commonly

Figure 35. A roadcut. Although the *Manual of Septic-Tank Practice* [15] notes, "Rock formations or other impervious strata should be at a depth greater than 4-feet below the bottom of trench or seepage pit," many local authorities look only for bedrocks and fail to note the qualifying work "impervious." Where bedrocks are fractured and will transmit rainwaters, it is technically only necessary to have enough topsoil into which to construct a shallow disposal field.

display rather permeable soils shallowly overlaying less permeable strata. Oftentimes, water-loving trees or plants will be seen where contours are concave (inwards) and form drainage courses for stormwaters. They would appear obvious hazards, but houses may be seen in drainage courses. And the convex (outwards) contours of hillsides can occasionally hold seasonal ground waters. Even at a very rapid speed of, say, 10 ft/day of ground water movement downslope, percolating waters will need several weeks to travel a few hundred feet to the bottom of a hill. Finding a perched water table on a steep hillside is not uncommon during a season of rains. Further, a technician should not be surprised to find seasonal ground waters at the top of a hill or at the top of a saddle between two hills.

Ground waters percolate in strange fashions oftentimes. In the Tehachapi Mountains, for example, ground waters may be found percolating in "fingers," as though along buried, shallow, and narrow drainageways. A deep pit can be dug with no signs of ground waters, but a disposal trench later will water-fill from intersecting a finger of ground water. Local knowledge is needed during soil explorations.

Roadways constituting dams to subsurface drainage are not limited to flatlands. Often a roadway passes across the foot of an ampitheatre-shaped geological unit in mountainous terrain. Those natural ampitheatres are watersheds which transmit drainages to a focal point below (another sad location for a house and disposal field). Further, geologically fault-blocked drainage from an "ampitheatre" at the side of a large meadow has been observed. Studies for large disposal fields, especially in mountainous terrain, should include geological input. What is a suitable open meadow to a household disposal field may be no more than a drainage swale to an uncommonly large disposal field.

It was mentioned that a driveway cut on a lot could bring effluents percolating downslope from above to daylight. The technician should look for permeability changes and slopes which would be candidates for such situations.

Slope, as such, is to the good. At the same time, however, it should be borne in mind that a backhoe, or whatever machine is most likely to construct disposal fields, cannot work safely on slopes too steep. A backhoe approaches capsizing as slopes become about 25 to 30% (about 15 to 17°). Sometimes it is local practice first to dig a terrace (or bench) along a contour, then to construct a disposal trench on that, and lastly to replace the soil spoil. That practice requires deeply permeable soils, otherwise the disposal field is installed into less permeable substrata or rock. In some areas where vegetation is slow to return, terracing is forbidden because of its ruinous aspects to the landscape.

SLOPES AND RUDIMENTARY WATER PHYSICS

In California, where, as said, strange things happen, one state authority, formally trained in other than septic-tank technology, promulgated belief that slopes were undesirable for disposal fields.* Some geology may be that way, but not most of California's. The reason offered was that percolating water would spring out of the hillside, the generally better success of disposal fields on hillsides notwithstanding. That misconception fails to account for the physical properties of water and displays an ignorance of geological conditions which will bring water to springs. A few remarks on those topics may help a technician evaluating good and bad features of a site.

Water can exist under tensile and compressive forces. Water in a barrel is under compressive forces, the deeper the water the greater the forces. If a valve is opened below, water will pour out in relief of compressive forces. Water in the compressive-force state forms a water table in soils. When percolating waters reach a much less permeable stratum, they can come to be under compressive forces. If that stratum goes to daylight, the water may also. Thus a spring is had.

Most people know that a pump cannot draw water up a pipe above about 34 ft, the column height at which atmospheric pressures are balanced. Botanists know that water in a very small pipe without air bubbles can be drawn higher than atmospheric pressure will push it: witness water in tall trees.† Water has tensile strength and a column of water can thus hang in a very small pipe.

Another way to view the tensile properties of water is to almost water-fill a drinking glass and set it in a saucer. Then fold absorbent paper to form a wick. Wet the paper and place one end into the water glass, long enough to reach the bottom. Drape the other end onto a saucer. A few hours later, the water will be in the saucer. To do so, water needs to be drawn uphill in the glass. Its tensile properties serve. The water does not fall to each side of the lip of the glass.

Water has surface tension. If a board is sloped, water can be sprayed

* The Point System was promulgated by staff of the California Water Resources Control Board during the middle 1970s. Backed by high authority, that system violated enough technology to fill volumes of scientific rebuttals.

† Vessels in tall trees containing gases frustrate explanations of water transport in solid columns under tensile stress. By overlapping notches, the writer has severed all vessels in some twigs over 40 ft high in Eucalyptus trees and found water transport unimpaired. Unheard, rare plant physiologists have theorized movement of water under tensile stress as cylinders within vessels, coexisting with gases. R. B. Krone has provided a conforming theory on the macromolecular structure of water [157].

onto the underside. The water will run down the underside and not fall off.

Water percolating downslope a hillside may not be in quantities enough to form a water table over a less permeable substratum. Those percolating waters will be under tensile stress. Capillarity in small soil pores "pulls" at the water. By no stretch of the imagination will those waters flow to daylight because of the steep slope.

A gopher, a burrowing animal in the West, is commonplace. He excavates passageways in soil to a larger hole for a nest. It does not rain inside of a gopher hole. Rainwaters stay in the soil and percolate around his excavations.

Slopes, as such, encourage drainage from disposal fields. The technician investigating a site should look favorably on slopes. He should at the same time, nevertheless, look for substrata capable of perching ground waters (compressive forces) and conducting them to daylight. Ordinarily, ground water situations are best displayed well after a season of rains begins. And, the technician should be aware of possible intersection of later construction with less permeable substrata downslope of contemplated disposal field sites.

On occasion, local authorities have asked whether or not a disposal field on a hillside would cause new springs or affect the existing ones. Most generally, disposal fields only add to percolation of rainwaters having pathways established by soils and geology. It would not ordinarily be expected that a new spring would develop. That would require a much increased use of a percolation pathway just on the verge of becoming a spring. Most likely, some waters might be added to existing springs. During a drier season, most plant life will welcome subsurface irrigation and withdraw waters percolating downslope from a disposal field.

SOIL STABILITY

Another aspect of slopes is soil stability. Landslides are unpleasantly commonplace in California's Coastal Range. Some counties require a professional report on the stability of a hillside building site. Geologists and soils engineers are rather disinclined to risk a multithousand-dollar lawsuit over a job of a few hundred dollars. Consequently, jobs tend to go to the more suicidally inclined professionals, or those that have little to lose. Unfortunately, California's legal system does not always encourage good practices.

Generally, landslides are caused by nature's forces, not new, and ongoing. As land erodes, part of that process is the bringing of large land masses towards sliding, perhaps for the first time. Landsliding is customarily triggered by heavy rains.

When considering a possible slide, commonly professionals view deep backhoe holes to determine whether or not movement, or creep, has taken place before. The possibility of a slide may or may not be related to past performances of the area. It would be well to consider the water a disposal field could offer to the hillside and then compare it to rainfall. Usually rainfall comes out far ahead. One also could gather history of landslides caused by disposal fields. One may look a very long time indeed. None are known to this writer.

LOCAL CAPABILITY

It is always tempting to solve a local problem with a new device, which "can work." But the answer to the question, "Will it work?" may be "no."

Local contractors must be capable and perhaps unusually willing to construct devices new to their locale. Still, there is a rather understandable tendency for contractors to overprice construction of devices new to their practice.

Without constant surveillance by the responsible designer, a contractor enters into considerable liability constructing a device new to his area [110].

Inspection services under best circumstances can only be a short visit. It is impractical in most localities to expect an authority to offer constant surveillance of construction of new devices.

The technician investigating a site with special devices in mind should account for local capability. That includes insistence of local authorities and financial backing of a client for the ongoing involvement of the consultant making recommendations. Very generally, new devices should not be advised unless unusually favorable conditions exist.

Mounds and evapotranspiration (ET) systems are new to many areas. Promulgated by the U.S. EPA, those devices are being installed without having been developed properly in new areas.

During a lecture on mounds, Professor Capon, from one of our northern universities, projected pictures of construction on a screen. When asked about his supervising presence, Capon remarked, "I just caught the contractor doing something wrong, one morning early." He further remarked, "We had to watch him all the time." Apparently unaware of the import of that experience, Professor Capon and his colleagues provided directions for using those mounds. Those directions have been disseminated without instructions to have constant surveillance of construction by an expert. Mounds have somehow served in some places, but certainly not in a good many others.

Locally Expected Wastewater Volumes

Expected wastewater volumes are a part of site evaluations. Ordinarily local authorities will mandate design considerations, but the designer must reckon with the real world. The site investigator must know what kind of water-using facilities are planned, and the designer should estimate wastewater volumes, even though local mandates are followed.

Commercial establishments vary enormously in water usage. Restaurants use much water and discharge much grease. They ordinarily need grease-interceptor tanks. A warehouse is relatively "dry." Estimates of typical wastewater flows from establishments are available [4,15].

Perhaps the most misleading authoritative guidelines advise design of disposal fields in accordance with the number of bedrooms a house is to have. Bedrooms do not generate wastewaters—people do. The designer will be closer to right if he considers a household producing a wastewater flow typical of his area, with a safety factor as well.

With some misgivings, water-saving devices are advised for consideration. There are many so-called water-saving devices available which actually result in increased usage of water. The reader is advised to read articles referenced [112,158]. Further, the water-saving movement belongs beside the "energy conversation" movement—mostly talk. The writer routinely suggests water-saving devices of quality to clients and authorities, but he feels like a member of a cult in doing so. Water-saving efforts have had very little genuine authoritative support.

The writer includes some 3.5-gal/flush toilets in foregoing remarks. Retrofitted devices of a same principle have too often resulted in double-flushing and consequently increased wastewater flows.

COMMENTS

A question has yet to be answered in many places: "Is the expected objective of subdivision work a demonstrated feasibility of septic-tank practices for each and every lot, or is it a feasibility of the subdivision as a whole with imperfections expected?" In some areas the former case is clearly demanded. But in other areas, the second situation is tacitly assumed, and it is expected that a few problem lots will one day be found. Unfortunately, the latter case leaves an engineer interested in staying in business necessarily staying within quite limited budgets. He cannot justify as much work as he would believe adequate, to clients who believe local authorities demand what is really needed. They often do not.

CHAPTER 12

SANITARY SURVEYS AND SURVIVAL CURVES OF SEPTIC-TANK SYSTEMS

Historically, the transition from private to public responsibilities for household wastewater management has been an unhappy affair. Always the issue of where to place responsibilities is confused with the tools themselves—septic-tank systems and sewers. Perhaps the advent of public entities charged with responsibility for all wastewater management—septic-tank systems or sewers—will be the end of the unhappy affair one day. Information published in 1975 [159] offers a tool to people in decision-making processes.

Occasionally one finds a customary septic-tank system which has served for fifty years or longer. Invariably when it reaches the end of its lifespan it is termed a failure, even if total neglect for half a century was the cause. Automobiles that need repairs, light globes that need replacement, knives that need sharpening—those are not regarded as failures. Still, the term failure is a part of septic-tank jargon, so that expression of prejudice will be used herein to communicate in familiar terms.

INTRODUCTION

It is not unusual to find an area, perhaps an older one, with more septic-tank problems than seem tolerable. Customarily, the local Health Department makes a survey of the area to find the percentage of failing septic-tank systems. Then the newspapers carry stories of serious health hazards and assert the need to sewer up. The populace splits into pro-sewer and pro-septic-tank camps, and the fight is on. There always seem to be two victims of the battle—the Health Department, which is attacked from several sides, and the poorer members of the community, who really can't afford the sewer costs. Worse yet, it is becoming increasingly obvious that a failing septic-tank system can be more of a public nuisance than a health hazard, and public sewers stimulate growth of an area to proportions planners and citizens may not have wanted [149,160].

From the two-decade investigations of existing septic-tank systems by the USPHS, the following conclusions might have been made:

1. If there be any great proportion of failures of septic-tank systems, it will happen in a year or two after construction.
2. Septic-tank systems have lifespans the lengths of which depend on several factors. Where soils, systems, usage, and various factors are reasonably consistent, the septic-tank systems have lifespans which behave as die-away phenomena. Curves of survival with years can be drawn.
3. Survival curves and related data could be used to predict lifespans of systems constructed in customary, same practices. Improved practices would produce a longer than usual lifespan.
4. Where septic-tank systems are installed over a period of years, there is no chance for mass failure or anything like it, unless of course, no one ever repaired the system.

A SURVEY OF SEPTIC-TANK SYSTEMS

In one county in California, there are two unincorporated communities, both with the same terrain, both with essentially the same septic-tank practices, and they are located near one another. The two communities differ in one quite important aspect—one has small, city-sized lots and the other has larger lots, mostly one acre or larger.

Small Lotsville

During the last few years, the County Health Department, Public Works Department, and other governmental agencies have applied great pressures on Small Lotsville to construct sewers, roads with curbs and gutters, and other things that those authorities regarded as better ways of life for the people in Small Lotsville. Also, Small Lotsville is right on the edge of a large city which would enjoy gobbling up Small Lotsville. After all, big cities thrive on population growth and taxes.

Unfortunately, the people in Small Lotsville did not want to be gobbled up. They liked their community the way it was. They also did not look forward to assessments of several thousands of dollars each. Some residents on fixed incomes or otherwise less wealthy would have had to sell their homes and leave the community. Other, wealthier residents, realized that sewers and other "improvements" meant even higher densities of development, and unwanted urbanization would be inevitable.

The engineers in the Public Works Department felt that Small Lotsville needed improved roads, gutters, and curbs. After all, it's hard to accommodate greater numbers of cars and greater speeds without extensive and expensive improvements. Still, the aesthetics of these engineers were not strong arguments. The strongest argument used for forcing the people in Small Lotsville to accommodate outside pressures was, of course, sewers to replace failing septic-tank systems.

The County Health Department made a survey of the septic-tank systems in Small Lotsville and found about 17% failures. With the political climate being what it had grown to be, the Health Department was placed in an unpleasant position. The health hazards were asserted critical. Newspaper articles would have one suspect that the people in Small Lotsville were going to drop off like flies, the existing good health of the community notwithstanding.

Another part of the picture was the belief that the soils in Small Lotsville were completely saturated with sewage from the septic-tank systems there. In reading newspapers, one could almost hear the feet in Small Lotsville going squish, squish through sewage-saturated soils. The people in Small Lotsville, however, still had the same garden-watering tasks they had always had in their droughty area.

Big Acres

The people in Big Acres, being close neighbors to Small Lotsville, watched the events in small Lotsville with concern. The people in Big Acres liked their community the way it was, and they did not wish to be gobbled up and urbanized. Realizing that soils and septic-tank practices were essentially identical in the neighboring communities, the citizens of Big Acres decided to look into their septic-tank situation to see how vulnerable they would be to a fate similar to that of Small Lotsville.

The Big Acres Homeowners Association performed a survey of septic-tank systems in their community. That survey can be used to illustrate the "how to do it" of survival curves. Most sanitarians and engineers do not know that survival curves describe septic-tank practices, because techniques for constructing survival curves are not a part of most literature.

Collection of Data

Surveyors of the USPHS reported that homeowners rarely knew much about their septic-tank systems and were emotionally reluctant to be surveyed. The writer's experience corroborates those observations. Based on his experience with homeowners, the writer offers the following guidelines:

1. Sanitary surveys should be performed by persons truly without strong interests in the outcome. Local Health Departments or other governmental bodies are usually viewed with suspicion, especially by a homeowner with septic-tank problems. Engineering firms able to design public sewerage systems have obvious conflicts of interest. Engineers, especially large sanitary engineering firms, should not survey septic-tank practices. Ideally, surveys should be conducted by knowledgeable persons, clearly believed by homeowners to have the homeowners' interests in mind [156].
2. Surveys made during hours when housewives are likely to be found at home alone should be performed by female personnel.
3. Surveyors should be genuinely friendly, be well trained in dealing with the public, be able to answer technical questions without offering advice, and be extremely patient. Homeowners frequently require considerable lengths of time or return visits before feeling secure enough to divulge perhaps unpleasant histories of sewage problems. Surveyors must be able to sense whether or not it is to a homeowner's advantage to retain information or to offer false information. Obtaining accurate data requires personal skills in dealing with people.

The survey of Big Acres was necessarily conducted by mail response with personal followups where needed. This procedure was considered adequate to provide a general overview, and engineering judgments or actions which would have required accurate and complete data were not then contemplated. Data collected described, among other things, the length of time the septic-tank system had served before being troublesome, if troublesome at all, and the nature of troubles.

CONSTRUCTING A SURVIVAL CURVE

As mentioned, techniques for constructing a septic-tank survival curve are not commonly known or available; thus a technique will be described.

In Table XV, each X in the Failed column indicates one septic-tank system that failed during the particular number of years indicated; for example, one system failed during the fifth year of use. An X in the Working column indicates a system which had successfully reached the indicated age and was still working; for example, four 5-year-old systems old had continued to work satisfactorily up through the time of the interview.

In Table XVI, Age Interval was chosen by looking at Table XV and deciding on a grouping of data intended to provide enough points to graph and not so many as to amount to an unnecessary number of scattered points. Arbitrarily, 5-year intervals were chosen. The Number of Installations was determined by counting all Failed and Working Xs within particular age intervals in Table XV. Then, adding the total Number of Installations yielded 84, heading the column Number Reach-

Table XV. Survival of Septic-Tank Systems In Big Acres—Data Tabulation

Years	Failed	Working
1	X	XX
3		X
4		X
5	X	XXXX
6		XXXXXX
7	X	XXX
8	XXX	XXXXX
9	X	X
10	XXXXXXX	XXX
11	X	XXX
12	X	XX
13		XXXX
14		X
15	X	XX
16	X	XXXX
17	X	XXXXX
18		X
19		XX
20	XX	XX
21		XX
22		X
23		X
24	X	XX
28	X	
30	X	X

Table XVI. Survival of Septic-Tank Systems in Big Acres—Calculations

Age Interval, Years	Number of Installations	Number Reaching Age Group	Number Failed	Quotient	Surviving %	Failed %
0–5	11	84	3	0.036	100	3.6
6–10	30	73	12	0.164	96.4	15.8
11–15	15	43	3	0.070	80.6	5.6
16–20	18	28	4	0.143	75	10.7
21–25	7	10	1	0.100	64.3	6.4
26–30	3	3	2	0.667	57.9	38.6

ing Age Group. The following number, 73, was calculated by subtracting 11 Number of Installations from 84; and 73 less 30 was 43, and so on. The Number Failed was determined by adding the Xs in the Failed column of Table XV, within the age groupings; for example, in the age interval 6–10 yr, 12 installations failed. Quotients (rounded off) are of Number Failed divided by Number Reaching Age Group. The Surviving column presents the percentage of installations surviving to reach Age Intervals; of course, the Surviving column begins with 100% of all systems. Then, multiplying the Quotient 0.036 by the corresponding 100% Surviving yielded 3.6% Failed. In turn, 3.6% was subtracted from 100%, which yielded 96.4% Surviving. Again in turn, 0.164 times 96.4 yielded 15.8, and that from 96.4 yielded 80.6, and so on.

Having constructed Table XVI, the graphing of data in Figure 36 was next. There, the percent Surviving was plotted against time. Matters were simplified by placing the 100% surviving point at time zero, the 96.4% surviving point at 5 years, and so on. Generally, survival curves are not straight lines, although in this example a straight line did fit best.

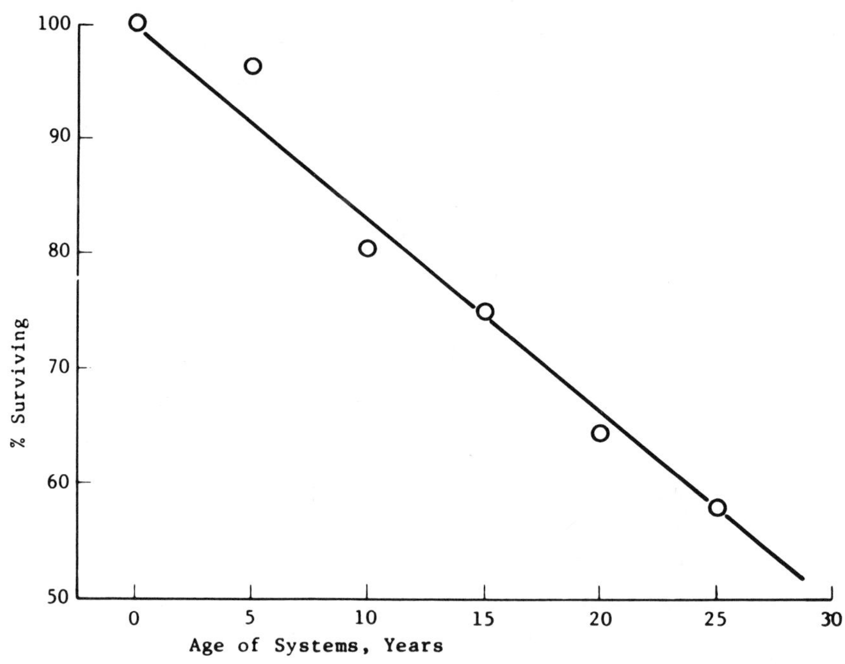

Figure 36. Survival of septic-tank systems in Big Acres.

MEANINGS OF DATA

In Table XV, 25 systems were noted Failed. Of that number, seven of the systems were not disposal field failures, because the septic tank settled and broke loose from the outlet tile; one system had a pump wear out which was replaced; and five of the systems had wooden tank covers rotted out and they needed new covers. It was tempting not to regard these systems as having failed, especially the one with only a pump that needed replacement. Still, they were left in the Failed column to compensate somewhat for the few homeowners who might have not reported troubles. A thorough survey should discover the nature of troubles so that criteria for failure could be sensibly made; such decisions are best postponed until an understanding of problems is had.

In looking at Figure 36, it can be seen that 50% of the systems survived almost 30 years. With such information, a decision on whether or not to continue with septic-tank practices could be sensibly made.

FALLACY OF CUSTOMARY SANITARY SURVEYS

Part of the survey of Big Acres included calculation of the "failure rate", which had been done by the County Health Department in Small Lotsville. Everyone was surprised to discover that the failure rate in Small Lotsville of 17% was essentially matched by 15% in Big Acres. Sanitarians of the County Health Department were aware of problems in Small Lotsville. Telephone reports of nuisances regularly came from there because when somebody's sewage daylighted, it ran onto someone else's property. In Big Acres when sewage daylighted, it seldom left a property before soaking back into the soils. Thus, the Health Department received few, if any, telephone reports of nuisances from Big Acres. Because of large lots, no one suspected problems in Big Acres. Indeed, so far as nuisances in Big Acres were concerned, there were few, if any problems.

Meaningless "Failure Rates"

Septic-tank systems not designed to serve indefinitely have lifespans, however short or long. When a system fails, the homeowner usually repairs it by adding more disposal trenches or the like. He does, however, take his time.

When a customary sanitary survey is made of an area and only the percentage of failures is noted, the data only tell us how many people need to repair [156]. Note in Table XVI that 25 of 84 systems had failed. That would have been a failure rate of about 30%, if no one ever repaired their system. Some did repair and the "failure rate" was reduced to 15%.

Failure percentages tell us nothing about lifespans. Only with measurements of survival curves could an engineer arrive at a valid cost comparison of community sewers vs continued septic-tank practices.

Further, without scientific input of septic-tank technology—not a part of most engineering curricula [32]—the benefits of improved subsurface disposal-field practices could not be assessed. More than one person believes that many federally funded sewerage projects have unintentionally and out of ignorance of septic-tank technology been an unneccessary waste, have urbanized reluctant communities, and have been a massive abuse of our less wealthy citizens [160].

CHAPTER 13

RESPONSIBILITY—PRIVATE AND PUBLIC

A septic-tank system requires very little maintenance. It rarely even gets that.

Essentially, septic-tank systems are predesigned by codes. The effort in a new situation is to see if the lot will tolerate the code system. If not, a lot which might have accommodated another kind of wastewater management system goes wanting. But, codes are necessary to local authorities who are neither empowered nor funded to provide individual designs. Only when properly empowered management entities are provided will septic-tank systems be individually designed.

HOMEOWNER MANAGEMENT

Most septic-tank systems have no moving parts. They are simple. And despite the extreme nuisance a disproportionately small number of systems have been, most have served trouble-free for decades.

The human brain responds to stimuli. Unless a threshold is crossed, nothing happens (assuming it does even then). A quiescent septic-tank system, innocuously buried and forgotten, historically has been an item of complete neglect. No one defends homeowner maintenance. It is manifestly inadequate. And when population densities are high, inadequate homeowner maintenance constitutes a need to change from private to public responsibility for wastewater management.

PRIVATE INDUSTRY MANAGEMENT

In recent years, private enterprise has successfully maintained individual wastewater disposal systems [46]. Requirements for success are:

1. authoritative support
2. economic feasibility
3. skilled personnel
4. almost trouble-free systems

Authoritative support includes many obvious aspects. Probably the most important is the granting of permission for private management to have its own designs for disposal fields which may be expected not to conform with local practices. Economic feasibility likewise has obvious aspects, but most critically there must be enough customers in an area to justify travels, etc. Skilled personnel are not easily found. They must be trained by the private enterprise itself. There is also a need for continuity of personnel for ongoing friendly relationships with customers.

Although almost trouble-free systems are needed, it is helpful if there are obvious duties for maintenance personnel. A septic-tank system with risers, valves to alternate fields, observation pipes—all appear more like something to do. A layman sees little to be done to a system buried in its entirety. That aspect of customary septic-tank systems, incidentally, is doubtless the largest factor in homeowner neglect. It is difficult for a homeowner to believe authoritative admonitions for maintenance when that same authority advises complete burial of all aspects of the system.

PUBLIC MANAGEMENT

The U.S. EPA has in recent years become interested in encouragement of public entities to manage on-site wastewater disposal systems. Other agencies have also become interested. Nevertheless, such entities have made small inroads into the vast collection of homeowner-neglected septic-tank systems. That was predicted [109]. The concept cannot become reality when artificially fostered by political interests. There is more grass-roots work yet to be done.

California has been the vanguard of the On-Site Wastewater Management District (more commonly termed Septic-Tank District), although California has no greater need than other states do [74,75]. The reason was a combination of three vital ingredients: firstly, a client willing to support a district in order to solve his problems (sometimes a small community defending itself against sewering schemes); secondly, a consultant aware of the need for ongoing responsibility to bring his recommendations to successful and maintained realities; and thirdly, a local authority willing to accept a subdivision for septic-tank practices—sometimes borderline in meeting local requirements—governed by a district.

As would be expected, efforts to bring the District concept to practical reality have had a variety of expressions. Districts vary from unfunded to funded, from cooperative agencies to essentially police districts, from awkwardly staffed to those skillfully guided, and from unsuccessful to successful.

Need for Continuity of Good Technical Direction

Human psychology has not changed much since ancient times. Biblical accountings manifestly display ancient understanding of having responsibility and authority vested in the same individual—the consequences of an act laid onto its instigator—the benefits of an act returned to its originator.

With the advent of the industrial revolution has come mass production of material goods and technicians. Material goods have interchangeable parts, and there has been an unfortunate tendency to regard technicians also as interchangeable parts.

Early engineers wrote of problems from a separation of authority from responsibility, a separation of the design engineer from maintenance of a small sewage treatment plant [91]. They described a 3-year history of maintenance difficulties and remarked, "The operation of the plant was manifestly a failure." Then, "At the end of this unsuccessful experience, the designing engineers were again retained to supervise the operation." Those writers went on to describe thereafter successful operations.

Experience with the variety of districts has displayed need for a consultant, recognizably proficient in the rare discipline of septic-tank technology, to be the same person to perform original site investigations, to offer recommendations for kinds of systems, to offer input to the arrangements for district involvement, to offer surveillance of construction of initial facilities, to offer training of the district's staff, and to serve the district until the project reaches a successful independence.

Through years of in-field experience the writer has had countless occasions to view construction of specially designed septic-tank systems. Without exception, the contractor needed and appreciated the presence of the designer in constant surveillance of all construction of the system. Constant surveillance of construction constitutes a rather large cost and the need for the consultant's time an unusual cost, but the need for the consultant's time is real.

As one might suspect, a most completely authorized district evolved out of peculiar circumstances—the happenstance of a distressed subdivision, a consultant with a district in mind as a solution to political problems, and local and regional authorities ready to accept the district concept in satisfaction of demands which could not be entirely met by physical conditions of the subdivision itself [74]. In other words, it helps if the subdivider is willing and financially capable of enticing or creating a public entity to accept responsibility for the design, inspection, maintenance, and monitoring of whatever is proposed. Local authorities must be convinced that the particular consultant's professional history uncommonly qualifies his capabilities and his willingness to see the project at hand to independence. Local authorities must be willing to grant approvals conditional to the foreseen involvement of the client, consultant, and responsible public entity.

The consultant must be willing to commit himself to a time-consuming program with long-term responsibilities. The client's financial strength must be respected, because the program about amounts to marriage. Fairness to the financially obligated client must be ascertained. A contract would be likely had, but the only good contract resides in the character of the people themselves.

A public entity must accept the program and be obliged to see it to completion. Possibly the entity would be the best vehicle for long-term financial arrangements. It must be controllable by technical interests favorable to the project, at least until the program is thoroughly a matter of routine. A private entity may serve, perhaps on a contract basis, to the public entity.

Pitfalls to Avoid

In two California counties, maintenance of septic tanks is a mandatory service administered to some homeowners. In the first county, a fee is collected at the time of inspection, and whether or not to have the tank cleaned at homeowner's expense is decided on-site by the inspecting authority. That program costs less than the program in the second county. There, the septic-tank is periodically pumped, need it or not. But higher overall costs are divided and collected monthly. The second, more expensive program is the more acceptable to all concerned. As said, the threshhold of attention must be crossed in the human mind. The first, less expensive program crossed that threshhold unfavorably.

In another California county, a district has essentially all authority needed to govern septic-tank practices. Nevertheless, that district has not been charged by ordinance with the responsibility for accomplishment. In consequence, the district has chosen to avoid responsibilities and simply serves in reflection of county practices. That is an example of failure to foresee need to couple authority with responsibility.

Another district began as a developmental tool. Rather than assume responsibility for periodic cleaning of septic tanks, the district was only to police homeowners to the task. Of course, the policing was not a welcomed chore, and it was not performed. When the day of reckoning came, homeowners resorted to political pressures to avoid having their financial and convenience threshholds crossed. The district would better have performed periodic cleaning services and billed homeowners as a small periodic tax with other taxes.

Rather than use skilled judgment with proper tools to ascertain sludge and scum levels (Volume II, page 98, Figure 17), some districts have chosen to advise pumpers following a table of allowable sludge accumulations, such as the *Manual of Septic-Tank Practice* [15] provides. No septic-tank pumper known to this writer possesses proper tools for measurements of scum and sludge. That suggestion is impractical. Only by trained district personnel with proper tools and record-keeping capability can tanks be cleaned when genuinely needed.

In one of our states, an engineer, Mr. Fleecem, wanted to take advantage of EPA financial benefits for districts. He enticed local authorities to sponsor his studies of an area. He found very minor problems, largely none. Still, Fleecem had decided at the onset to initiate a district, so he advised the local people to support an unneeded office full of staff doing unneeded work. Where building lots are quite large and septic-tank practices quite successful, a district would very likely constitute an economic waste.

CHAPTER 14

SOME FIELD APPLICATIONS

Through years of consulting, many problems have been solved with septic-tank systems which did not meet specifics of local codes, but which met code provisions giving local authorities authorization to accept systems they believed would conform with the intentions of codes and surpass specifics.

INVERTED SAND FILTER— AN ARTIFICIAL SOILS SYSTEM

Principle

In 1961 [62], disposal trench models were constructed in which soil sidewalls and bottoms each had a 1-in. layer of fine sand over it, a 1-in. layer of coarse sand over that, and a central filling of roofing gravel. Injected effluents necessarily passed out of the distribution pipe, and in turn through the roofing gravel, coarse sand, fine sand, and into the soil. Those trench models functioned better than did companion trenches without sand layers between same gravels and soils.

In a separate experiment, the same arrangement of gravel, sands, and soil as was had in the trench models was placed in a clear glass permeameter. Accumulations of black deposits were seen at each interface where particle sizes of layers changed. Black deposits of ferrous sulfide had been found to be a good indicator of soil clogging from sewage spreading [48]. It was observed that clogging at each sand interface did not amount to total clogging of that permeameter equal to the clogging at a single interface which was provided in a companion permeameter filled with gravel and soil only.

Providing a multiplicity of interfaces between progressively finer layers of sands to infiltrating wastewaters, each managing a proportion of filtration and treatment, was superior to the customary provision of a single soil interface to infiltrating wastewaters in disposal fields. Also, use of a multiplicity of interfaces would make it possible to construct a relatively compact device to remove the clogging capacity of a septic-tank effluent before injecting it into a natural soils system. And, confining clogging to a replaceable man-made device, in situations where space was limited on a lot and no room for expansion was available, would preserve a natural soils system.

Without good design criteria when opportunities were had for field applications, inverted sand filters were overdesigned. Bagged sands graded for sand blasting, aquarium use, etc., were available. And, whereas vertical layers of different sand sizes could be assembled in a laboratory, field constructions seemed limited to horizontal layers.

A Slowly Permeable Soil

In St. Helena, California, a house was to be constructed on a hilltop, facing a gentle slope. The architect/builder faced an old orchard soil which was very slowly permeable. He wanted something innovative for his new home.

At about the hilltop, a notch was cut directly into the hill. The notch was about a dozen feet wide, and the bottom was slightly sloped downwards. Towards the front of the notch, cutting was continued into a sweeping turn around the face of the hill. That formed a terrace with its bottom continuing along a slight downslope from the rear of the notch.

The terrace was terminated as the cut tapered at the end at roughly 50 ft from where it turned away from the notch into the hill. An about 1-ft-deep layer of river sand was laid from the rear of the notch to the end of the terrace.

Within the notch, a horizontal frame of 2- × 4-in. lumber was used to make a 12- × 16-ft rectangle. A fine sand was spread within. Progressively, same-sized frames were laid and sandfilled with coarser sands than below. Thus, four layers of sands were constructed, each 4 in. thick, graded from fine to coarse (30, 16, 12, 8 mesh). Above, a 4-in. layer of medium gravel contained perforated distribution pipes 3-in. in diameter. They were laid in a fork-like fashion with tines 4 ft apart. Coarse sand and a medium sand were then placed in 2-in. layers each over the gravel. Particles were sized such that each layer would not infilter one another. River sand was used to cover the box and to fill between the sides of the box and the soil walls of the notch. Soil was then used to cover the excavated notch to grade. Only the terrace of river sand remained exposed. The end result was a buried box containing an inverted sand filter overlaying a tongue of river sand emerging from the buried box and running out onto a terrace with a gentle downslope. Septic-tank effluents would necessarily percolate downwards through the sand filter to soil. Then effluents would tend to percolate out under, and through the layer of river sand until they either seeped into the slowly permeable soils of the terrace or were evapotranspired.

The septic tank was followed with a 150-gal dosing tank with a pump set to discharge all contents at a time. Whereas dosing of disposal fields seems pointless, dosing sand filters is advantageous [48,52].

That system, built in the fall of 1966, served a house occupancy of four until 1970, three until 1972, and two persons to 1980 when inquiry was made. It had been believed dosing was needed, but the pump failed in 1978 and had not been replaced. Apparently the two-person wastewater flows did not overload the then anaerobic filter.

The tongue of river sand on the terrace has served subsurface irrigation of a pear tree and much grass. No odors have been noticed.

A Sandstone Hill

In late 1967, a frustrated homebuilder had a lot on top of a hill of sandstone in Scott's Valley. The sandstone was not much fractured. Percolation tests in holes dug with mallet and chisel had yielded poor results.

An inverted sand filter was suggested. Sewage clogging would be limited to the filter while the poor percolation capacity of native sandstone would be protected and maintained.

With much effort, a backhoe dug a trench 1 ft wide and 3 ft deep, and in a rectangle 10 × 20 ft. Over a 2-ft-deep filling of gravel in the trenches, a barrier was placed. Then a redwood box (10 × 20 ft) 4 ft high was constructed overhead. About a foot of fill sand was placed within the box. Then an inverted sand filter was constructed. It in turn was topped with soil to a few inches under the top of the box.

The filter was as described foregoing, except for the distribution system. Instead of using ordinary perforated pipes, a pressurized system of 1.5-in.-diameter plastic pipes was laid about 1 ft on centers. Each pipe had two rows of holes drilled to ⅛ in. diameter each. The rows were on each side of the pipe, and holes were spaced in rows about 6 in. apart. When water tested before gravel covering, the activated pump caused water from each hole to form an arc about 3 in. outwards.

A 1200-gal septic tank was followed by a 750-gal dosing tank with a pump. Water-using habits of two adults caused the discharge of the liquid contents about each 5 to 6 days. When the pump went on, the owner reported a very slight odor which lasted no longer than a couple of minutes. A person had to be close to the filter to notice the odors.

Having functioned trouble-free to date, an overflow system has not seen service. The trench containing the pipe connecting the dosing tank to the distribution system in the filter had been gravel packed around the pipe. The excavation containing both the dosing and the septic tanks had been likewise gravel packed, fitted with a barrier, and soil filled to grade.

Any effluents which could not have been managed by the system of trenches under the sand filter were to overflow down the pipe trench to the larger, tank-containing excavation. An observation pipe within that excavation has been used by the homeowner with which he ascertained no liquid levels to date.

A Fractured Bedrock

A restaurant and shelter at the top of Squaw Peak needed disposal facilities for wastewaters. The candidate area for a disposal field was a fractured metamorphic bedrock. Inasmuch as infiltration would necessarily be into small fractures, it was judged that clogging of the tight fractures might proceed rapidly if septic-tank effluents reached them directly. Whereas soils may routinely have 50% porosity, the fractured bedrock must have had a porosity less than 1%.

Because of freezing temperatures, disposal needed to be subsurface. A bed consisting of 6000 ft^2 was constructed with a Caterpillar D-9 with a smoking ripper. Then an inverted sand filter containing approximately the same sand layers described before was constructed therein. Dosing of pressure distribution pipes with ¼-in.-diameter holes was achieved by hydraulic head from a tower uphill.

That inverted sand filter has served since winter of 1977–78. Customarily it receives extremely heavy ski season use and very little summer use.

An Unconsolidated Sandy Soil

At two localities—one in Scott's Valley, California, and one in Port Charlotte, Florida—inverted sand filters have been constructed to serve single-family residences. Each filter was placed directly onto sandy soils, and there was no constructed excavation for infiltrating effluents below the filters. Permeabilities were such that effluents percolating through the filters could be transmitted through the natural sands below.

Both filters have served 4 or 5 years to date. Both were provided with sampling ports below the filter for monitoring effluents. Still, no data have yet been reported on treatment achieved.

A Smaller Inverted Filter

During FHA-sponsored septic-tank system studies at the University of California, research efforts were much assisted by George Tchobanoglous. His interest in residentially scaled wastewater disposal systems persists as he now serves on the faculty of the School of Engineering, University of California, Davis.

In 1967, Tchobanoglous assisted with a solution to a residential problem in Los Altos Hills. He recommended an inverted sand filter which was constructed in a rectangular box offering 60 ft^2 in horizontal area (6 ft × 10 ft). It was served by a 540-gal dosing tank. Two persons used the filter, and it must have been loaded at 9 gal/ft^2 on an about 4-day dosing cycle. Unfortunately from the data-collecting point of view, the filter was abandoned in favor of a public sewer connection by the owners after a few months of operation, and long-term performance history was not to be had.

Comments

Inverted sand filters require much special attention in design and during construction. Whether or not dosing on a several-days-per-dose schedule is needed is not known. The experiences have been empirical.

Sand filters and recirculating sand filters have been used successfully in residential applications. They are simpler in construction than inverted filters and likely less expensive [11]. The foregoing experiences have not been presented to favor a kind of filter. They are to exemplify solutions to problems where customary disposal fields were less feasible.

Compact, maintenance-free, sand filters are man-made and therefore replaceable. They make possible the use of natural conditions of certain sites where sewage clogging of porous media would be most undesirable. Many lots are too small to accommodate ordinary systems with room for repairs.

SANDS

Unconsolidated sands may collapse into excavations. In disposal fields, they will migrate into ordinarily coarse gravel fillings. An angle of repose of a sand could be on the order of 30°.

Disposal Beds—Horizontal Surfaces

If a wide disposal bed in sand is to be constructed, the designer can plan on sloping sidewalls. Those sidewalls and the flat bottom surfaces can serve as infiltration surfaces.

Ordinarily, excavated sands will be used to backfill to grade. Whereas soils may consolidate and not filter down through a decomposing barrier, sands will likely not consolidate and they will filter down through a decomposing barrier into gravel fillings below. A barrier of a permanent material should separate the top of the gravel filling of a bed from sand backfill overhead. There are several plastic cloths available which will serve. Also, barriers of natural materials can be constructed by broadcasting fine gravel over a coarser gravel filling. A coarse sand is broadcast over the finer gravel. After a finer sand is broadcast to cover the prior material, native soils are used to backfill to grade. A designer should first collect the available materials and place one on top of the other in a jar to be sure one does not infilter the other. Although one might believe this approach requires much time, it is convenient and practical where properly graded materials are available. The results are perfectly fitted, permanent, and porous barriers.

Disposal Pits—Vertical Surfaces

In 1975, a system of disposal pits was designed to serve Rio Bravo Lodge, Bakersfield, California. Because the lodge had an unusual design, authorities of the Kern County's Division of Environmental Health, the developer of Rio Bravo Ranch, the design engineer Mr. Charles Karoly, and the consultant (the writer) all demanded constant surveillance of construction by the design engineer or his technician. Soon it was found that a bucket auger could not construct the pits as planned because a stratum of unconsolidated, collapsing sand was encountered. Being present, the engineer altered the design to accommodate construction practicalities. As stated earlier, the constant surveillance by the design engineer is needed and increased quality of construction will result.

Each of 28 pits was constructed 36 in. in diameter and with about 17 ft of useful sidewall height. Rather than customary gravel (drain rock), ½-in. gravel was used to support the integrity of the excavations.

Two alternate fields of 14 disposal pits each were serially distributed by a horizontal solid pipe. Over each pit was a cross-fitting with a solid riser to above grade and a perforated pipe to the pit bottom. An elevated jumper pipe connected the solid distribution pipes leading to the two disposal fields, at a place just beyond the 3-way valve following the septic tank. Thus, if one field ever filled, effluents would flow to the alternate. Such a circumstance was thus ensured but not expected.

Designed to manage 6000 gpd, after 7 years of operation, about 75% of the infiltrative surfaces has yet to be exposed to effluents. Further, monitoring indicates that the ½-in. gravel has adequately supported the vertical sidewalls in strata of clean, coarse sand, of medium sand, and fine sand.

Vee Trench—Sloped Surfaces

In 1981, Mr. George Nickel foresaw need for more subsurface wastewater disposal facilities to serve his large, high-quality development of Rio Bravo Ranch, Bakersfield, California. The writer was assigned the task of recommending facilities to manage perhaps 500,000 gpd.

During soils investigations, very large deposits of unconsolidated sands were found [161]. It was decided to provide interceptor tanks and a public pressure sewer, designed to function by gravity from most lots. A single, large disposal field was to follow a pumping station at a point lower than connections to the pressure sewer and the disposal field. Then, in cooperation with Ricks, Taylor & Meyer, Inc., of Bakersfield, and Weatherby Associates of Jackson, plans were provided which contractors submitted were within construction practicalities.

In 1982, 7 ft of topsoil overburden were removed and stockpiled. The freshly exposed sands collapsed when dug, as studies predicted, and the contractor found he could not construct trenches as he had planned. Redesigns were provided by the consultant and engineer, their being present in constant surveillance of construction.

A vee-shaped trench was dug 11 ft deep (18 ft below original grade) and 24 ft wide at its top. As the slightly moist sands dried, they began falling down the 92% sloping sidewalls in seeking their more gentle angle of repose. Before much movement took place, ¾-in. gravel was unloaded into the trench end. Unloading more gravel upon in-place gravel, a front end loader pushed the gravel in a cascading tongue down the centerline of the trench. The effect was gravel falling into place rather than scouring the delicate sidewalls of sand (Figure 37).

A single perforated distribution pipe was laid down the centerline of the trench brimfilled with gravel. That pipe was then covered with gravel in a long mound. A porous, plastic cloth was spread as a barrier, over which stockpiled topsoils were replaced to grade.

The contemplated system was to be quite large. Because it was known that a competent public entity, the Olcese Water District, was to be responsible for maintenance of the sytem, it was decided to construct three trenches amounting to a total of 800 lineal ft as a first sector,

Figure 37. A vee trench was dug into slowly collapsing sands. Gravels were unloaded onto gravel from one end, rather than onto delicate sidewalls. Then gravels were pushed into place from the center, rather than scouring the sidewalls.

expected to serve 500–600 homes. As development proceeds and wastewater flows are managed by the first sector, its performance will be studied by the consultant to the Olcese Water District, and results will be used to predict a final design for the entire 500,000-gpd system. Actual performance would constitute a far better design criterion than would any known practical program of soils measurements.

Where local authorities can abandon belief that only vertical sidewalls

Figure 37., continued.

of pits and horizontal surfaces of trenches (or beds) are actually serviceable, sloping surfaces of a vee trench can be accepted as a geometry suitable for excavations in unconsolidated sands.

SILTS

Soils lacking cohesive strengths will migrate into percolation test holes. They will also migrate into gravel-filled disposal fields, often clear up to distribution pipes. Whereas sand grains are large and heavy enough to be held in place by medium to fine gravel, finer silts may questionably be held in place with pea gravel.

Vertical, supporting layers of sands do not lend themselves to construc-

Figure 37., continued.

tion convenience. It would seem that quite broad trenches or beds would best serve. A designer is advised to provide a permanent barrier, capable of excluding fine silt particles. As with sands, backfills of silts cannot be expected to consolidate such as to remain in place over a decomposing barrier.

MINIPITS—MAXIMIZING ASSIMILATION

It is not unusual to look at a lot with a slowly permeable soil and wish there was a way to spread wastewaters evenly over the lot and still have subsurface disposal. Larry Baker of Weatherby Associates, Jackson,

California, has come as close to utilizing the entire assimilative capacity a site has to offer as would seem practical.

In a county in sourthern California, authorities and a contractor failed to recognize the folly of placing a disposal bed at the focal point of a drainage swale [110]. Looking from the disposal bed uphill, the lot spread out much as a side of a funnel would. Soils were very slowly permeable and overlay even less permeable subsoils. The only hope of a proper repair was use of the uppermost contour and thereby maximizing assimilation of downsloping percolating effluents before the focal point of the funnel-like swale would be reached. At the same time, a disposal trench around the top of the lot beside the paved roadway would have been impractical to construct. Authorities forbade that approach.

First, the house was fitted with water-saving devices [158]. Then in repair, a septic tank uphill was followed by a sump below with a dosing siphon. Effluents were dosed onto a low-maintenance sand filter. In turn, clear sand-filter effluents were collected in a sump with a pump. Then clear effluents were pumped through a filter used to protect the drip irrigation system following that.

At convenient places around the upper periphery of the lot, a series of disposal pits were each dug 1 ft in diameter and 3–4 ft below finished grade. Each contained ¾-in. gravel under a barrier and soil backfill. At grade and extending down into the gravel filling was a capped 4-in. solid plastic pipe. The pressurized drip-irrigation pipe traversed through the center of that capped pipe. Within it, a drip emitter was placed. Thus the small pit received clear wastewaters through serviceable drip emitters.

Similar arrangements of small, separated units of disposal fields served by drip emitters have been used successfully by Baker in other areas. They are dependent on solid-free effluents, homeowner maintenance, quality electrical controls, and knowledgeable placement of the elements over the site such as to maximize what assimilative capacity there is. Generally, very slowly permeable soils are the case and hydraulic loading rates must be low, depending on the specific circumstances.

POTTERS

There are more people making pottery than one might realize. That craft yields wastewaters containing clays. Those particles are so small as to stay in suspension in a septic tank. They can be expected to pass through. Clays are certainly detrimental to a disposal field. They clog soils. And, being minerals, clays cannot degrade and they clog permanently.

Two neighbors are full-time potters. They were advised to have their studio sink and drain separate of other household plumbing. That studio drain left the building foundation at below frost line, about 30 in. deep in Santa Fe, and then entered a gravel-filled pit about 4 × 4 ft square and perhaps 5 ft deep. The soils were sandy, had a percolation rate of about 12 in./hr (5 min/in.), and had a reaction of about pH 8.9 from caliche.

After about 6 months, the clay-containing wastewaters had clogged approximately 60 ft^2 of bottom and sidewall surfaces of the pit. Clogging was permanent, so a soil-exhaustion rate of 60 ft^2/yr/potter was had.

Being rural and on about 2.5 acres, an open trench with sidewalls sloped for safety would seem an acceptable device. Rainfall is only about 5–10 in./yr and would not periodically water-fill the trench. Those potters plan to enlarge the pit into a trench about 4 × 15 ft, and have an easily accessible end slope. Gravel will be piled over the wastewater drain at one end such as to protect it from freezing. The remainder of the open 5-ft-deep trench will be available for infiltration.

That open portion of the trench should serve for about 2 years before it becomes clogged. Then it could be permitted to dry (a rapid occurrence in the desert sun at Santa Fe's 7000-ft elevation). When dry, the clay accumulations could be raked up and removed.

Essentially, the particular potters' problem is being managed by separating studio wastewaters from the residential septic-tank system, which would be rapidly ruined by the permanently clogging clay wastes. Those separated clay-laden wastewaters are being filtered by rather permeable soils. In a less favorable soil situation and where surface discharge would be unacceptable, other filtration arrangements must be devised by local improvisation.

VERY SMALL WASTEWATER VOLUMES IN HEAVILY GOVERNED RURAL SETTINGS

A few years ago, several people identifiably members of a subculture acquired rural lands in a few of California's counties. Those people chose to build their own homes which were often comparatively small, artistically designed rather than standard in dimensions and such, and generally had none of the code-demanded conveniences—electricity, gas, telephones, and pressurized water supplies. Being beyond roadways, many people did quite nicely with about 10 gal of hand-carried fresh water daily.

Conformance, voluntary or compulsory, is a human trait. So, members of the subculture conformed to their very conservative way of life. But, that way of life was at variance with the values of conformance of the major culture. As would be expected, the major culture became rather oppressive and a part of that oppression was the forceful and extreme application of codes.* Ordinarily, use of a pit privy and throwing about 10 gpd onto a few acres of land would not excite anyone. But some local authorities demanded full-scale septic-tank systems as though to serve households with pressurized water supplies. At the same time, pit privies and compost privies became subject of a state-supported research project. Incredibly, many thousands of dollars were spent by health authorities to discover whether or not a privy could be safely used.

The writer would fearlessly throw 10 gal or so of wastewaters from a kitchen sink, or wherever, out of his back door onto his 6 acres. Nevertheless, where authorities demand and police subsurface disposal of even such small wastewater volumes, but will permit a system proportionate to realistically expected volumes, a suggested arrangement is presented in Figure 38.

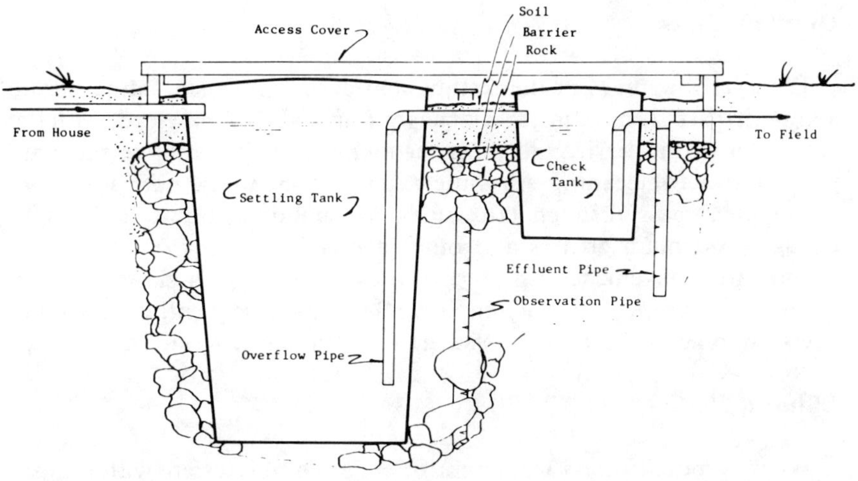

Figure 38. System for treatment and disposition of very low volumes of domestic wastewaters.

*The Health Officer of one of California's northern counties joined the sheriff in raiding a subculture community. The Health Officer remarked, "Those hippies bought some of our land and they act as if they owned it!"

Settling Tank

Wastewaters coming from the house (likely a kitchen sink), enter the settling tank. There, greases and other debris are to be settled and stored between cleanings. The settling tank consists of a plastic garbage can, roughly 30 gal in capacity. Such cans have lids fitting tightly enough to exclude insects, but not so tight as to require external relief venting.

Inasmuch as most wastewaters are likely to have considerable quantities of cooking greases, the bottom of the overflow pipe is shown at about 80% under the liquid surface. Adjustments of the length of the outflow pipe may need to be made—up if more sludges are had, or down if mostly scum is had.

Check Tank

The check tank should have a minimum of about 5 gal capacity. It should be made of plastic and have a lid.

Overflow Pipes

The overflow pipes in the settling and check tanks should be plastic and perhaps 1–2 in. in inside diameter. Joints should be glued, with the exception of the horizontal pipes leading out of the tanks to facilitate pipe removal for cleaning. Absolute watertightness would not be essential where pipes pass through tank sides. They should, however, be tight enough to contain greases and other debris.

Only for convenience, the overflow pipes have been shown located opposite inlet pipes. In function, it would be inconsequential if inlet and overflow pipes were closer together, or even side by side, if wished.

Effluent Pipe

Where ground waters would not be expected to interfere with infiltration, and the excavation for the tanks has been prepared as shown in Figure 38, effluents from the check tank could be distributed by dropping them into rock backfillings through an effluent pipe glued into the bottom of a tee. The plastic effluent pipe should have holes (about ½ in.) drilled into it, but perhaps it might be easier to cut notches into the pipe with a knife or saw. No holes or notches are to be placed where the pipe is to pass through soil backfill, a barrier, or layers of finer gravel capable of falling into the notch.

The effluent pipe is to ensure open passage through soil backfill and into coarser gravel or stones below. Once well into those coarser materials, the pipe needs to continue no further. Although drawn as a straight piece in Figure 38, the effluent pipe can function if contorted downwards.

Observation Pipe

A straight, vertical pipe should extend from just under the access cover to the bottom of the excavation. The pipe should be 3–4 in. in diameter and be provided with a removable cap on top. Notches (or holes) should be cut into the pipe within the coarser gravel or stones, but no cuts or holes should be in the part passing upwards through soil or other finer backfill materials. The spacing of the notches is not critical, their function being the assurance of water levels in the pipe representing true levels of whatever waters occur in the excavation.

Backfill Soil, Barrier, and Rock

The excavation should be large and deep enough to contain the tanks and pipes planned for use, with a few inches of space to spare at the sides and bottom of the excavation. Gravel or small stones should be placed under the settling tank for support and to allow effluents to infilter native soil beneath. Over that gravel, coarse gravel or even large stones might be placed as shown. If large stones are used, smaller stones on top would close large voids and support a barrier.

The barrier could be an impermeable plastic sheet, a permeable or impermeable building paper, or perhaps straw. The barrier is intended to keep soil backfill (as shown) from infiltering gravel below. The soil backfill is to seal off odors and about 6 in. would do.

Gradients

Being free of large solid debris, wastewaters will flow from a house to a nearby settling tank without much concern for slope of the pipe. It is only needed that the water can fall. The overflow pipes in the settling and check tanks should be about level, although the check tank could be lower perchance it is located in a separate excavation for some reason.

Subsurface Disposal Field

If the excavation drawn in Figure 38 is, or is believed to be, insufficient for infiltration, a separate subsurface disposal field could serve. Although directions for sizing such fields have been presented in this volume, most likely the homeowner would choose to construct a trench to his own convenience.

Maintenance

Whenever sizeable quantities of scum and sludge appear in the check tank, both the check tank and the settling tank should be cleaned.

The purpose of the observation pipe is to ascertain liquid levels within the excavation. When they become excessively high, additional infiltration capacity should be considered.

Should odors become a nuisance, lids of the settling tank and check tank can be sealed into place with silicone. In such case, however, the homeowner must consider the venting of those tanks so that flows from the house drain are not air-locked.

If liquid levels within the excavation are observed rather high, care must be taken in cleaning the settling and check tanks. They could be floated out of position. It may be necessary first to pump waters from the excavation out of the observation pipe before cleaning the settling and check tanks. As said, such a ponding may indicate need for additonal infiltration capactiy.

ODORS

The primary concern people have with their septic-tank system is odor. And, oftentimes odors are emitted from house roof vents which ordinarily communicate directly with the septic tank. One septic-tank pumper manufactures an activated-charcoal filter for roof vents, the market being there. A more direct way to exclude odors from house vents is to cap the inlet tee in the septic tank or exchange it with an ell.

When risers are had and odors emanate through the customarily poorly fitted lids on septic tanks, silicone can be used to seal lids. They can be readily lifted when needed and resealed.

Ordinarily, soils of disposal fields sorb odors. But sometimes odors are very noticeable over an uncommonly large disposal field, such as would serve a community. When a designer is contemplating a very large disposal field, he should design, but not necessarily construct, a facility for deodorizing effluents before they are discharged to the disposal field. Oxidation of sulfides of hydrogen is the objective. Chlorine and ozone are commonly available oxidants.

REFERENCES

1. Enferadi, K. M. "A Field Evaluation of the Waterless Toilet as an Alternative to the Failing Soil Absorption System," paper presented at the Eighth Annual National Sanitation Foundation Conference, Ann Arbor, MI (19–22 October 1981).
2. Olivieri, A. W., and R. C. Cooper. "Public Health Aspects of Onsite Systems," paper presented at the Eighth Annual National Sanitation Foundation Conference, Ann Arbor, MI (19–22 October 1981).
3. Bailey, J. R., R. J. Benoit, J. L. Dodson, J. M. Robb and H. Wallman. *A Study of Flow Reduction and Treatment of Waste Water from Households*, Report to USDI—Federal Water Pollution Control Administration, General Dynamics, Electric Boat Division, Groton, CN (December 1969).
4. Metcalf & Eddy, Inc. (G. Tchobanoglous et al.). *Wastewater Engineering*, (New York, NY: McGraw-Hill Book Co. 1972).
5. Winneberger, J. T., R. Laak, W. D. Hypes and K. Y. Chen. *Manual of Grey Water Treatment Practice*, (Ann Arbor, MI: Ann Arbor Science Publishers, Inc., 1976).
6. Laak, R. *Wastewater Engineering Design for Unsewered Areas*, (Ann Arbor, MI: Ann Arbor Science Publishers, Inc., 1980).
7. *Audel's Plumbers and Steam Fitters Guide #2*, Theo. Audel & Co., Publishers (1925).
8. Smith, L. P.O. Box 1256, Pahoa, HI. Personal Communication (1982).
9. Winneberger, J. H., A. B. Menar and P. H. McGauhey. *A Study of Methods of Preventing Failure of Septic-Tank Percolation Fields, Third Annual Report*, Sanitary Engineering Research Laboratory Report No. 63-9, University of California, Berkeley (22 December 1963).
10. Winneberger, J. T. "Report of Screening of Flocculants," Report to R. E. Wann, Church & Dwight Co., Inc., New York, NY (1 September 1976).
11. Bowne, W. C., Consulting Engineer, Rt. #3, Box 4032, Roseburg, OR. Personal Communication (1982).
12. Schmidt, H. E., Consulting Engineer, 641 West 53rd Street, Hialeah, FL. Personal communication (1982).
13. Winneberger, J. T. "Watertight Septic Tanks?" *Land and Water* 25:16 (1981).

14. Bendixen, T. W., R. E. Thomas and J. B. Coulter. *Study of Seepage Pits*, Report to FHA, USPHS, Robert A. Taft Sanitary Engineering Center, Cincinnati, OH (1 May 1963).
15. Joint Committee on Rural Sanitation and staff of the USPHS. *Manual of Septic-Tank Practice*, Public Health Service Pub. No. 526, U.S. Government Printing Office, Washington, DC, Revised (1967).
16. *Design Manual, Onsite Wastewater Treatment and Disposal Systems*, U.S. Environmental Protection Agency, Office of Water Program Operations, Office of Research and Development, Municipal Environmental Research Laboratory, Cincinnati, OH (October 1980).
17. McGauhey, P. H., and J. H. Winneberger. *Final Report on A Study of Methods of Preventing Failure of Septic-Tank Percolation Systems*, Sanitary Engineering Research Laboratory Report No. 65-17, University of California, Berkeley, (31 October 1965).
18. McGauhey, P. H., and J. H. Winneberger. *A Study of Methods of Preventing Failure of Septic Tank Percolation Systems*, U.S. Department of Housing and Urban Development, U.S. Government Printing Office, Washington, DC (October 1967).
19. Coulter, J. B., and T. W. Bendixen. *Effectiveness of the Distribution Box* Report to FHA, USPHS, Robert A. Taft Sanitary Engineering Center, Cincinnati, OH (19 February 1958).
20. Coulter, J. B., Chief of Suburban Sanitation Studies, U.S. Public Health Service. Letter to John G. Morris, Chief Sanitary Engineer, Lake County Health Department, Waukegan, IL (14 April 1959).
21. Winneberger, J. H., and R. B. Krone. "Report of a Field Examination of Marginal Soils for Use in Disposal of Household Wastes," Report to the Technical Studies Section of the FHA, Sanitary Engineering Research Laboratory, University of California (1 July 1962).
22. Ryon, H. *Notes on the Design of Sewage Disposal Works with Special Reference to Small Installations*, Private Publication, Albany, NY (1928).
23. Winneberger, J. T. "Suitability of On-Site Wastewater Disposal Systems Along California's Coast," Occasional Paper No. 1, Office of Appropriate Technology, Sacramento, CA (June 1976).
24. Olivieri, A. W., and R. J. Roche. *Minimum Guidelines for the Control of Individual Wastewater Treatment & Disposal Systems*, California Regional Water Quality Control Board, San Francisco Bay Region, Oakland, CA (April 1979).
25. Ludwig, H. F., and G. W. Ludwig. "Improved Soil Percolation Test," *Water and Sew. Works* 97(12):513–16 (1950).
26. Ludwig, H. F., W. D. Ward, W. T. O'Leary and E. Pearl. "Equilibrium Percolation Test," *Water Sew. Works* 97(12):513–516 (1950).
27. Art's Machine Shop, Harrison at Oregon Trail, American Falls, ID 83211.
28. Cedergren, H. R. *Seepage, Drainage, and Flownets*, (New York, NY: John Wiley & Sons, Inc., 1967).
29. Winneberger, J. T. "Correlation of Three Techniques for Determining Soil Permeability," *J. Environ. Health* 37(2):108–118 (1974).
30. McGauhey, P. H., G. T. Orlob and J. H. Winneberger. *A Study of the Biological Aspects of Failure of Septic Tank Percolation Systems, First Progress Report*, Sanitary Engineering Research Laboratory, University of California, Berkeley (31 December 1958).

31. Winneberger, J. "*Practical Uses of New Septic-Tank Technology*," *J. Environ. Health* 30(3):250–262 (1967).
32. Winneberger, J. T., and J. W. Klock. *Current and Recommended Practices for Subsurface Waste Water Disposal Systems in Arizona*, ERC-R-73014, College of Engineering Sciences, Arizona State University, Tempe, AZ (July 1973).
33. Bendixen, T. W., M. Berk, J. P. Sheehy and S. R. Weibel. *Studies on Household Sewage Disposal Problems, Part II*, U.S. Public Health Service, Robert A. Taft Sanitary Engineering Center, Cincinnati, OH (1950).
34. Winneberger, J. T. *Keene Center, Feasibility of Subsurface Disposal of Wastewaters, Kern County, California* Report to Donald E. Ward, C.E., Rickett, Ward & Delmarter, Bakersfield, CA (15 September 1982).
35. Winneberger, J. T. *Studies of Feasibility of Subsurface Disposal of Wastewaters at Corte Madera Ranch, Portola Valley, California, Report I*, Report to Aries Enterprises, Inc., Redwood City, CA (5 November 1979).
36. Winneberger, J. T. *Studies of Feasibility of Subsurface Disposal of Wastewaters at Corte Madera Ranch, Portola Valley, California, Report I*, Report to Aries Enterprises, inc., Redwood City, CA (20 February 1980).
37. USDA Soil Conservation Service. *Soil Service Manual*, USDA Handbook No. 18, U.S. Government Printing Office, Washington, DC (August 1951).
38. Weibel, S. R., T. W. Bendixen and J. B. Coulter. *Studies on Household Sewage Disposal Systems, Part III*, U.S. Public Health Service, Robert A. Taft Sanitary Engineering Center, Cincinnati, OH: (1954).
39. Duley, F. L. "Surface Factors Affecting the Rate of Intake of Water by Soils, *Proc. Soil Sci. Soc. Am.* 4:60–64 (1939).
40. Duley, F. L., and L. L. Kelly. "The Effect of Soil Type, Slope and Surface Conditions on Intake of Water," Nebraska Agricultural Experimental Station Research Bulletin 112 (1939).
41. Winneberger, J. H., and P. H. McGauhey. *A Study of Methods of Preventing Failure of Septic-Tank Percolation Fields, Fourth Annual Report*, Sanitary Engineering Research Laboratory Report No. 65-16, Report to FHA, University of California, Berkeley (31 October 1965).
42. Winneberger, J. *The Engineering Feasibility of Individual Residence Sewage Disposal Installations*, Report to Geldermann, Inc., Danville, CA (1 August 1965).
43. Smith, L. *The Sun for a Hat* (Philadelphia: Dorrance & Co., 1976) and personal correspondence (1975).
44. Winneberger, J. T. *Report to General Development Utilities, Inc., Septic-Tank Practices, Critique and Suggestions for Improvements*, Report to Harold E. Schmidt, President, General Development Utilities, Inc., Miami, FL (15 November 1977).
45. "Regulations Governing the Approval of Individual Sewage Disposal Systems," and "Soil Percolation Testing Procedure," San Mateo County Department of Public Health and Welfare, Office of Environmental Health, County Office Building, 590 Hamilton Street, Redwood City, CA.
46. Winneberger, J. T., W. L. Mellen and F. L. Matthew. "Maintenance by Private Industry of On-Site Wastewater Disposal Systems, The Septic System, and A Simple Formula for Successful Wastewater Systems," *On-Site Waste Management, Volume VIII*, (Findlay, OH, Hancor, Inc., 1980).

47. *Uniform Plumbing Code*, 1976 ed., International Association of Plumbing and Mechanical Officials, Los Angeles, CA (1975).
48. Winneberger, J. H., L. Francis, S. A. Klein and P. H. McGauhey. *Biological Aspects of Failure of Septic-Tank Percolation Systems, Final Report*, Sanitary Engineering Research Laboratory, University of California, Berkeley (31 August 1960).
49. Winneberger, J. H., A. B. Menar and P. H. McGauhey. *A Study of Methods of Preventing Failure of Septic-Tank Percolation Fields, Second Annual Report*, Sanitary Engineering Research Laboratory, University of California, Berkeley, (22 December 1962).
50. Laak, R. *The Effect of Aerobic and Anaerobic Household Sewage Pretreatment on Seepage Beds*. PhD, University of Toronto, Canada (1966) Doctoral thesis.
51. Winneberger, J. T. "Separation of Water-Borne Wastes," in *Manual of Grey Water Treatment Practice*, J. T. Winneberger, Ed. (Ann Arbor, MI: Ann Arbor Science Publishers, Inc., 1976), pp. 68–78.
52. Otis, R. J. University of Wisconsin, Madison. Personal Communication (4 February 1977).
53. Wood, P. K. "Report, Demonstration Project, Survey of Sanitation District No. 3, Boyd County, Kentucky," Department for Natural Resources and Environmental Protection, Bureau of Environmental Protection, Frankfort, KY (10–13 April 1978).
54. Winterer, E. V. "Percolation of Water Through Soils," *California Agricultural Experimental Station Annual Report from July 1, 1920 to June 30, 1921*, 160–161, University of California, Berkeley (1922).
55. Winterer, E. V. "Percolation of Water Through Soils," *California Agricultural Experimental Station Annual Report from July 1, 1922 to June 30, 1923*, 234–237, University of California, Berkeley (1923).
56. Winterkorn, H. F. "Mechanism of Water Attack on Dry Cohesive Soil Systems," *Soil Sci.* 54(4):259–273 (1942).
57. Christiansen, J. E. "Some Permeability Characteristics of Saline and Alkali Soils," *Agric. Eng.* 28:147–150 (1942).
58. Christiansen, J. E. "Effect of Entrapped Air on the Permeability of Soils," *Soil Sci.* 58:355–365(1944).
59. Allison, L. E. "Effect of Microorganisms on Permeability of Soil under Prolonged Submergence," *Soil Sci.* 63:439–450 (1947).
60. Christiansen, L. E., and O. C. Magistad. "Report for 1944, Laboratory Phase of Cooperative Water-Spreading Study," USDA Regional Salinity Lab., Bureau of Plant Industry, Soils and Agricultural Engineering, Washington, DC (1944).
61. McGauhey, P. H., G. T. Orlob and J. H. Winneberger. *A Study of the Biological Aspects of Failure of Septic-Tank Percolation Systems, Second Progress Report*, Sanitary Engineering Research Laboratory, University of California, Berkeley (1 July 1959).
62. Winneberger, J. H., W. I. Saad and P. H. McGauhey. *A Study of Methods of Preventing Failure of Septic-Tank Percolation Fields, First Annual Report*, Sanitary Engineering Research Laboratory, University of California, Berkeley (22 December 1961).
63. Otis, R. J., W. C. Boyle, J. C. Converse and E. J. Tyler. *On-Site Disposal of Small Wastewater Flows*, Prepared for the Environmental Protection Agency Technology Transfer, Madison, WI (1977).

64. Coulter, J. B., T. W. Bendixen and A. B. Edwards. Report to the FHA, *Study of Seepage Beds, Part I, Knox County, Tennessee*, U.S. Public Health Service, Robert A. Taft Sanitary Engineering Center, Cincinnati, OH (14 February 1960).
65. Coulter, J. B., T. W. Bendixen and A. B. Edwards. Report to the FHA, *Study of Seepage Beds*, U.S. Public Health Service, Robert A. Taft Sanitary Engineering Center, Cincinnati, OH (15 December 1960).
66. Federick, J. C. "Soil Percolation Rates," *Public Works* 83(7):46–48,89–90 (1952).
67. Federick, J. C. "Solving Disposal Problems in Unsewered Areas," *Sewage Works Engineering* 19(6):292 (1952).
68. Winneberger, J. T. "Winneberger's Law," *J. Environ. Health* 34(1):58–61 (1971).
69. Weibel, S. R., C. P. Straub and J. R. Thoman. *Studies on Household Sewage Disposal Systems, Part I*, U.S. Public Health Service, Environmental Health Center, Cincinnati, OH (1949).
70. Orlob, G. T., and R. G. Butler. *An Investigation of Sewage Spreading on Five California Soils*, Sanitary Engineering Research Laboratory Technical Bulletin 12, University of California, Berkeley (1955).
71. Orlob, G. T., and R. B. Krone. *Movement of Coliform Bacteria Through Porous Media, Final Report*, Sanitary Engineering Research Laboratory, University of California, Berkeley (30 November 1956).
72. Winneberger, J. T. "The Principle of Alternation of Subsurface Wastewater Disposal Fields," *On-Site Waste Management, Volume V* (Findlay, OH: Hancor, Inc., 1976).
73. Winneberger, J. T. "The Alternate Bed System," *Proc. Ohio Home Sewage Disposal Conference*, Cooperative Extension Service, Ohio State University, Columbus (29–31 January 1973).
74. Winneberger, J. T., and W. H. Anderman, Jr. "Public Mangement of Septic-Tank Systems is a Practical Method of Maintenance," *J. Environ. Health* 35(2):145–146 (1972).
75. Winneberger, J. T., and J. A. Burgel. "On-Site Wastewater Management Districts," *On-Site Waste Managment, Volume VI* (Findlay, OH: Hancor, Inc., 1977).
76. Talley, L. R. Director of Environmental Health, Santa Cruz County Environmental Health Service, 701 Ocean St., Fourth Floor, Santa Cruz, California 95060. Personal Communication (1982).
77. Muskat, M. *The Flow of Homogeneous Fluids Through Porous Media*, (Ann Arbor, MI: J. W. Edwards, Inc., 1946).
78. Martin, J. P. "Microorganisms and Soil Aggregation: I. Origin and Nature of Some of the Aggregating Substances," *Soil Sci.* 59:163–174 (1945).
79. Martin, J. P. "Microorganisms and Soil Aggregation: II. Influence of Bacterial Polysaccharides on Soil Structure," *Soil Sci.* 61:157–166 (1946).
80. Bliss, E. S., and C. E. Johnson. "Some Factors Involved in Ground-Water Replenishment," *Trans. Am. Geophys. Union* 33(4):547–558 (1952).
81. "Diverter Valve," Hancor, Inc., P.O. Box 1047, Findlay, OH 45840.
82. "Diversion Valve," Franklin Research, 4009 Linden St., Oakland, CA 94608.
83. Huberty, M. R. "Compaction in Cultivated Soils," *Trans. Am. Geophys. Union*, 896–899 (1944).
84. Kozlova, L. I. "First Results of Micro-Cinematographic Study of Deforma-

tion of Soils Due to Percolation," *Izvestiya Naucho-Issledovatelskovo Instituta Hydrotechniki* 10:219–30 (1933), Abstract in *The Engineering Index*, 1014 (1933).

85. Kozlova, L. I. "Experimental Study of Motion of Fine Grained Sand Through Pores of Coarser Material," *Izvestiya Naucho-Issledovatelskovo Instituta Hydrotechniki* 14:210–227 (1934), Abstract in *The Engineering Index,* 999 (1935).
86. Lowdermilk, W. C. "Influence of Forest Litter on Run-off, Percolation, and Erosion," *J. Forestry* 28:474–490 (1930).
87. Healy, K. A., and R. Laak. *Factors Affecting the Percolation Test*, C.E. 72–51, University of Connecticut, Storrs, Connecticut (April 1972).
88. Laak, R. School of Engineering, Box U-37, Universtiy of Connecticut, Storrs, Connecticut 06268, Letter to J.T. Winneberger (24 September 1982).
89. Frank, L.C., and C. P. Rhynus. *Studies of Methods for the Treatment and Disposal of Sewage—The Treatment of Sewage From Single Houses and Small Communities*, Treasury Department, U.S. Public Health Service, Public Health Bulletin No. 101, December 1919, Washington DC, U.S. Government Printing Office (1920).
90. *Rules and Regulations for the Preparation and Submission of Plans for Systems of Sewerage and Sewage and Waste Disposal*, Bulletin No. 1, New York State Department of Health, Albany, NY (1940).
91. Metcalf, L., and H. P. Eddy, *American Sewerage Practice—Volume III, Disposal of Sewage*, 3rd ed., (New York: McGraw-Hill Book Company, 1935).
92. Joint Committee on Rural Sanitation and staff of the USPHS. *Manual of Septic-Tank Practice*, Public Health Service Pub. No. 526, Washington, DC, U.S. Government Printing Office, first printing (1957).
93. Birnbaum, A. "Energy Economics in Drain Tube and Drain Tile Production," Paper presented 1975 Winter Meeting of American Society of Agricultural Engineers, Chicago, IL (15–18 December 1975).
94. Knor, M. M., *Evaluation of Septic Tank Effluent Distribution*, Independent Undergraduate Research Project, University of Connecticut, Storrs (1973).
95. "Guidelines for Installation of Septic Tank Systems," Arizona State Department of Health Services, Engineering Bulletin No. 12 (May 1974).
96. Rouman, G. L., Dept. of Environmental Quality Program, Ferris State College, Big Rapids, MI. Personal communication (4 April 1974).
97. Crates, F. J. Alternative Wastewater Management Consultant, 13067 Twp. Rd. 204, Findlay, OH. Personal communication (November 1982).
98. Crates, F. J. "On-Site Waste Management: Past, Present & Future," *Land and Water* (January 1980).
99. Crates, F. J. "Management—A Many Splendored Word," prepared for the Onsite Waste Management Conference, Ann Arbor, MI. Hancor, Inc., P.O. Box 1047, Findlay, OH 45840 (1980).
100. Darcy, H. *Les fontaines publiques de la ville de Dijon* (1856).
101. *Alternate Sewage Manual*, Section of Plumbing & Fire Protection Systems, Wisconsin Department of Health and Social Services, Madison, WI (June 1975).
102. Kohler, M. A., T. J. Nordenson and D. R. Baker. *Evaporation Maps for the United States*, U.S. Department of Commerce, Weather Bureau, Technical Paper No. 37, U.S. Government Printing Office, Washington, DC (1959).

103. Pruitt, W. O., F. J. Lourence and S. von Oettingen. "Water Use by Crops as Affected by Climate and Plant Factors," California Agriculture (October 1972).
104. Morton, F. I. "Potential Evaporation and River Basin Evaporation," *Proc. ASCE* 91(HY6):67–97 (1965).
105. Pruitt, W. O. *Procedures for Estimating Crop Water Requirements for Use in Water Allocations and for Improvement of Irrigation Efficiency*, Water Resources Center, University of California, Davis (1964–1966).
106. Weaver, J. E., and F. E. Clements. *Plant Ecology*, 2nd ed., (New York: McGraw-Hill Book Company, 1938).
107. Copley, J. C., Ed., "Flood Control," *Oil-Power* 27(1):7 Socony-Vacuum (1952).
108. Marsh, A. W. "Single Tree Potential Water Use," *Irrigation and Drainage Letter*, No. 24, Agricultural Extension Service, University of California, Riverside (November 1965).
109. Winneberger, J. T. "A Consultant's Overview of On-Site Needs," *National Conference on Less Costly Wastewater Treatment Systems for Small Communities*, Reston, Virginia, U.S. EPA (12–14 April 1977).
110. Winneberger, J. T. "The Contractor—Fallguy for Poor Septic-Tank Codes and Administration, and Fallguy-Elect for the U.S. EPA's Efforts to Help On-Site Sewage Management," *Land and Water* 25:12–13 (1981).
111. Winneberger, J. T. "Using Water-Saving Devices to Prevent or Correct a Troublesome Septic-Tank System," *Land and Water* 26:25–26 (1982).
112. Baker, L. Weatherby Associates, Jackson, CA. Personal communication.
113. Winneberger, J. T. *Pollution in the Colusa National Wildlife Refuge and the Neighboring Environment*, Report to City of Colusa and California Regional Water Quality Control Board, Central Valley Region (7 March 1971).
114. Winneberger, J. T. *Health Hazards of Sewage Ponds in the Pacific Flyway*, Rept. to City of Colusa and California Regional Water Quality Control Board, Central Valley Region (12 April 1971).
115. Ward, P. Bureau of Sanitary Engineering, California State Department of Public Health. Personal communication (1971).
116. Greenberg, A. E. Chief, Sanitation and Radiation Laboratory, California State Department of Public Health. Personal communication (1971).
117. Kinnicutt, L. P., C.-E. A. Winslow and R. W. Pratt. *Sewage Disposal*, (New York: John Wiley & Sons, Inc., 1910).
118. Metcalf, L., and H. P. Eddy. *American Sewerage Practice, Vol. III, Disposal of Sewage*, (New York: McGraw-Hill Book Company, 1916).
119. Calabro, J. F. *Microbiology of Septage*, Doctoral Thesis, University of Connecticut, Storrs, Connecticut (1971).
120. Johnson, J. H., J. E. Fields and W. A. Darlington, "Removing Viruses from Water by Polyelectrolytes," *Nature* 213 (5077):665–667 (1967).
121. Parsons, F. *Incidence of Virus in Ground-Water Receiving Septic Tank Effluent*, MS Thesis, University of Miami, FL (1973).
122. Pitt, W. A. J., Jr., H. C. Mattraw, Jr. and H. Klein. *Ground-Water Quality in Selected Areas Serviced by Septic-Tanks, Dade County, Florida*, U.S. Geological Survey, Open File Report 75-607, Miami, FL (1975).
123. Krone, R. B. "The Movement of Disease Producing Organisms Through Soils," Presented at Symposium on the Use of Municipal Sewage Effluent for Irrigation, Louisiana Polytechnic Institute, Ruston (30 July 1968).

124. Eliassen, R., W. Ryan, W. Drewry, P. Kruger and G. Tchobanoglous. "Studies on the Movement of Viruses With Groundwater," Water Quality Control Research Laboratory, Stanford University, Palo Alto, CA (1967).
125. Hodgkinson, C. *Removal of Coliform Bacteria from Sewage by Percolation Through Soil*, Sanitary Engineering Research Laboratory, University of California, Berkeley (15 May 1955).
126. Orlob, G. T., and R. B. Krone. *Final Report on Movement of Coliform Bacteria Through Porous Media*, Sanitary Engineering Research Laboratory, University of California, Berkeley (30 November 1960).
127. Orlob, G. T. *Report on the Engineering Feasibility of Individual Residence Sewage Disposal Installations For the Rancho San Geronimo Area, Marin County, California*, Water Resources Engineers, Inc., Berkeley (October 1960).
128. Romero, J. C. "The Movement of Bacteria and Viruses Through Porous Media," *Ground Water* 8(2):37–48 (1970).
129. Gary, M., R. McAfee, Jr., and C. L. Wolf. *Glossary of Geology*, American Geological Institute, Washington, DC (1972).
130. Millon, E. R. *Water Pollution, Red Feather Lakes Area, Colorado*, MS Thesis, Colorado State University, Ft. Collins, CO (May 1970).
131. Waltz, J. P. Colorado State University, Ft. Collins, CO., Personal communications (1973).
132. Laak, R. *Nitrogen and Phosphorus Removal in a Septic Tank and Lagoon*, Experimental Investigation of Nitrogen and Phosphorus Removal at the University of Connecticut Research Station, University of Connecticut, Storrs, CN (July 1974).
133. Morrison, S. M., and M. J. Allen. *Bacterial Movement Through Fractured Bedrock*, Environmental Resources Center, Completion Report Series No. 32, Colorado State University, Ft. Collins, CO. (July 1972).
134. Waltz, J. P. "Methods of Geologic Evaluation of Pollution Potential at Mountain Homesites," *Ground Water* 10(1) (1972).
135. Waltz, J. P. "How Safe is Mountain Well Water?" CSU-Research, Colorado State University, Ft. Collins, CO (1973).
136. Cooper, R. C. "Waste Water Management and Infectious Disease. I.: Disease Agents and Indicators," *J. Environ. Health* 37(3):217–224 (1974).
137. Cooper, R. C. "Waste Water Management and Infectious Disease. II.: Impact of Waste Water Treatment." *J. Environ. Health* 37(4):342–350 (1975).
138. *Microbiological Content of Domestic Wastewaters Used for Recreational Purposes*, State Water Quality Control Board Pub. No. 32, Sacramento, CA (1965).
139. Merrell, J. C., W. F. Jopling, W. F. Bott, A. Katko, and H. E. Pintler. *The Santee Recreation Project, Santee, California, Final Report*, U.S. Department of the Interior, Federal Water Pollution Control Administration, WP-20-7, Cincinnati, OH (1967).
140. Winneberger, J. H., W. D. McMillan, R. F. Murphy and K. O'Brien. *Quality Control of Recreational Lakes*, Report to United States Land, Inc., Palo Alto, California (18 November 1968).
141. Winneberger, J. T. *Nitrogen, Public Health, and the Environment, Some Tools for Critical Thought*, (Ann Arbor, MI: Ann Arbor Science Publishers, Inc., 1982).

142. Wert, S. "Performance of Soil Absorption Fields Following Sand Filters, Two Case Studies." In preparation, Northwest Soil Consulting, Roseburg, Oregon (1982).
143. Sawhney, B. L., and J. L. Starr. "Movement of phosphorus from a septic system drainfield," *J. Water Poll. Control Fed.* 49(11):2238–2242 (1977).
144. Sawhney, B. L. The Connecticut Agricultural Experiment Station, New Haven. Personal communication (13 April 1978).
145. Sprague, R. W. *The Ecological Significance of Boron*, U.S. Borax Research Corporation, Anaheim, CA (1972).
146. Fireman, M., and H. E. Hayward. "Irrigation Water and Saline and Alkali Soils," *The Yearbook of Agriculture 1955, Water. The United States Department of Agriculture*, U.S. Government Printing Office, Washington, DC (1955).
147. U.S. Department of Commerce, Bureau of the Census, 1970 Census of Housing, Detailed Housing Characteristics, Arizona (April 1972).
148. Winneberger, J. T. Unpublished data (1974).
149. *Interceptor Sewers and Suburban Sprawl: The Impact of Construction Grants on Residential Land Use, Volume I, Analysis*, Prepared for Council on Environmental Quality, Urban Systems Research & Engineering, Inc., Cambridge, MA (31 July 1974).
150. Prickett, R. J. and R. C. Cooper. *Water Borne Viral Disease—A Complete Bibliography on Enteric Viruses in Water*, School of Public Health, University of California, Berkeley (1968).
151. Day, P. R. "Report of the Committee on Physical Analyses," *Soil Sci. Soc. Am. Proc.* 20(2):167–169 (1954–1955).
152. Krebs, R. D., and J. H. Hunter. *Judgment of Soil Features for Residential Sewage Disposal*, Report to FHA, Virginia Polytechnic Institute, Blacksburg VA (August 1961).
153. *Evaluation of Soils and Use of Soil Surveys for Engineering Purposes in Urban Development*, FHA No. 723, Work of Krebs and Hunter, compiled and edited by the Technical Studies Staff, Architectural Standards Division, Federal Housing Administration, Washington, DC (October 1963).
154. McMillan, W. D. Personal communication (1968).
155. Ford, H. W. Personal communication (1982).
156. *Septic-Tank Practices in Bolinas, California—The Human Ecology of Bolinas Mesa.* A report to Bolinas Community Public Utility District, Bolinas, CA: Warshall & Farnsworth and Associates (May 1973).
157. Krone, R. B. *A Theory of the Molecular Structure of Water*, Doctoral Thesis, University of California, Berkeley (1962).
158. Baker, L. K. *The Impact of Water Conservation on Onsite Wastewater Management*, Weatherby Associates, Inc., Jackson, Calif; paper presented at the Seventh National Conference on Individual Onsite Wastewater Systems, Ann Arbor, MI (23–25 September 1980).
159. Winneberger, J. T. "Sanitary Surveys and Survival Curves of Septic-Tank Systems," *J. Environ. Health* 38(1):36–39 (1975).
160. Warshall, P. *Septic-Tank Practice, A Guide to the Conservation and Re-Use of Household Wastewater*, (Garden City, NY: Doubleday Anchor, 1979).
161. Winneberger, J. T. *Subsurface Disposal of Wastewaters at Rio Bravo Ranch, Bakersfield, California*; Report to Mel McColloch, President, Olcese Water District, Bakersfield, CA (9 July 1981).

INDEX